Springer Theses

Recognizing Outstanding Ph.D. Research

Aims and Scope

The series "Springer Theses" brings together a selection of the very best Ph.D. theses from around the world and across the physical sciences. Nominated and endorsed by two recognized specialists, each published volume has been selected for its scientific excellence and the high impact of its contents for the pertinent field of research. For greater accessibility to non-specialists, the published versions include an extended introduction, as well as a foreword by the student's supervisor explaining the special relevance of the work for the field. As a whole, the series will provide a valuable resource both for newcomers to the research fields described, and for other scientists seeking detailed background information on special questions. Finally, it provides an accredited documentation of the valuable contributions made by today's younger generation of scientists.

Theses are accepted into the series by invited nomination only and must fulfill all of the following criteria

- They must be written in good English.
- The topic should fall within the confines of Chemistry, Physics, Earth Sciences, Engineering and related interdisciplinary fields such as Materials, Nanoscience, Chemical Engineering, Complex Systems and Biophysics.
- The work reported in the thesis must represent a significant scientific advance.
- If the thesis includes previously published material, permission to reproduce this must be gained from the respective copyright holder.
- They must have been examined and passed during the 12 months prior to nomination.
- Each thesis should include a foreword by the supervisor outlining the significance of its content.
- The theses should have a clearly defined structure including an introduction accessible to scientists not expert in that particular field.

More information about this series at http://www.springer.com/series/8790

Joseph Weston

Numerical Methods for Time-Resolved Quantum Nanoelectronics

Doctoral Thesis accepted by
the University of Grenoble, France

 Springer

Author
Dr. Joseph Weston
Qutech
Delft University of Technology
Delft
The Netherlands

Supervisor
Dr. Xavier Waintal
Institute of Nanoscience and Cryogenics
CEA Grenoble
Grenoble
France

ISSN 2190-5053 ISSN 2190-5061 (electronic)
Springer Theses
ISBN 978-3-319-87615-3 ISBN 978-3-319-63691-7 (eBook)
DOI 10.1007/978-3-319-63691-7

Printed on acid-free paper

This Springer imprint is published by Springer Nature
The registered company is Springer International Publishing AG
The registered company address is: Gewerbestrasse 11, 6330 Cham, Switzerland

Supervisor's Foreword

The development of the field of quantum nanoelectronics has always been strongly interlinked with microelectronics technology, in particular through the possibilities offered in clean rooms to design increasingly small samples made of a large variety of materials. A more recent trend originates from the progress made in dealing with microwave frequencies. Indeed, the temperature reached in a typical quantum nanoelectronics experiments lies around 20 mK which translates into 0.4 GHz. Introducing microwaves in the 10 GHz range into the experiment therefore opens up the possibility to study, for the first time, the quantum dynamics of these systems. The field started with closed systems (superconducting circuits, quantum dots), which are rapidly evolving toward quantum engineering. A new generation of experiments now targets open systems where excitations can propagate. There, one can design experiments that are the electronic analog of what has been developed in quantum optics. Various key components have already been demonstrated (coherent single electron sources, beam splitters) while the remaining ones (single electron detectors) are under intense development. There is little doubt that the study of quantum dynamics of fermionic systems will get increasingly amenable to experiments in the next few years. Likewise, the theory for this physics is well established and the problems can, at least, be formulated very neatly. There is however one crucial ingredient that has been lacking: the ability to actually solve these theoretical problems. Analytical solutions can be obtained in some regimes but, unlike the stationary case, there was until recently no numerical method fast enough to address these problems computationally. This is where the Ph.D. work of Joseph Weston fits in.

Computational quantum transport in d.c. is now a very mature field. Since the entire physics takes place close to the Fermi energy, solving a stationary quantum transport problem boils down to solving the single electron Schrödinger equation at the Fermi energy (not necessarily a trivial task for an open system). As soon as one start to consider time dependent systems, such as the propagation of the excitation created by a voltage pulse, one is dealing with a very different category of problem; the system is now inelastic (its energy changes as it absorbs or emit "microwave photons" into the environment) and one crucially needs to enforce the Pauli principle (an electron cannot be excited to an already occupied state) making the problem effectively many-body again. The main achievement of the Ph.D. work of Joseph is the development of a new numerical technique that can deal with systems with thousands of orbitals (our record is currently at 10^5) and/or very long simulation time (currently around 10^5 in units of the inverse Fermi energy) several orders of magnitude faster than the previous generation of numerical techniques. The Ph.D. work of Joseph has made possible the study of many interesting physical phenomena that were simply inaccessible before. Beside the technique itself, Joseph has designed a concrete software implementation that we hope to release to the public soon under an open source license, in order to allow the community to easily benefit from these developments. We hope that this software, "tkwant", which inherits many features from the Kwant package that we have been developing for some years, will be useful to many.

The reader will find that the technique can be used for a wide variety of physics that includes (so far) the propagation of voltage pulses in semi-conducting heterostructures, quantum Hall effect, Josephson junctions, Majorana fermions, magnetic materials, Floquet topological insulators and, we hope, many more that we did not consider so far.

Grenoble, France   Dr. Xavier Waintal
June 2017

Abstract

Recent technical progress in the field of quantum nanoelectronics has lead to exciting new experiments involving coherent single electron sources. When quantum electronic devices are manipulated on time scales shorter than the characteristic time of flight of electrons through the device, a whole class of *conceptually new* possibilities become available. In order to treat such physical situations, corresponding advances in numerical techniques and their software implementation are required both as a tool to aid understanding, and also to help when designing the next generation of experiments in this domain.

Recent advances in numerical methods have lead to techniques for which the computation time scales linearly with the system volume, but as the *square* of the simulation time desired. This is particularly problematic for cases where the characteristic dwell time of electrons in the central device is much longer than the ballistic time of flight. Here, we propose an improvement to an existing *wave-function based* algorithm for treating time-resolved quantum transport that scales linearly in both the system volume *and* desired simulation time. We use this technique to study a number of interesting physical cases. In particular we find that the application of a *train* of voltage pulses to an electronic interferometer can be used to *stabilise* the dynamical modification of the interference that was recently proposed. We use this to perform spectroscopy on Majorana and Andreev resonances in hybrid superconductor-nanowire structures.

The numerical algorithms are implemented as an extension to the KWANT quantum transport software. This implementation is used for all the numerical results presented here, in addition to other work, covering a wide variety of physical applications: quantum Hall effect, Floquet topological insulators, Fabry-Perot interferometers and superconducting junctions.

Acknowledgements

First and foremost I would like to thank my supervisor Xavier Waintal for his constant support during these three years. Doing a Ph.D. is not easy, and his open-door policy, combined with his infinite patience, has served as an anchor for me in during the times when I was most frustrated.

I would also like to thank Valerio Olevano, Julia Meyer and Hugues Pothier for being in my Ph.D. defence committee, and David Carpentier and Michael Wimmer for, in addition, reviewing the this thesis.

A large part of what has made the last three years unique is the people with whom I shared the experience on a daily basis. Vladimir Maryasin and Benoît Gaury (my predecessor) were my office mates from the beginning, and the physics and philosophy discussions we shared during the first two years were massively important to me. I will always remember Benoît as someone to whom the mathematics was always clearer than "fuzzy" explanations, and who helped me to formulate my ideas in more precise terms. I also spent good times with other members of the lab, most often over a beer and a pizza at the end of the week in the *Brasserie du Carré*; in particular the humour of Alexander Grimm, Alexandre Artaud and Lars Elster leaves me with particularly fond memories. In the same vein I would like to thank Christoph Groth, for being the person with whom I could always count on for lively discussions, mostly related to more geeky topics.

Finally I would like to thank my family. My parents Dave and Cathy have shown me unwavering support, even when they perhaps didn't agree with some of my decisions. Also to my sister, Sara, thank you for being one of the only people to whom I can speak frankly. Last of all I would of course like to thank my amazing partner Carolina for being there day after day; the end of this Ph.D. marks the beginning of our life together.

Contents

Chapter 1
Overview

This thesis deals with the problem of simulating and understanding transport of electrons in quantum devices when the device is subjected to time-dependent perturbations such as pulses of voltage on contacts or electrostatically coupled gates, electromagnetic radiation etc. This thesis consists of two main parts. The first part is dedicated to the advances in theoretical techniques and their implementation as numerical algorithms that allow us to perform simulations where the time to solution scales linearly in the system volume and maximum (simulation) time required. This is an improvement over the present state of the art, which scales linearly in system volume, but as the *square* of the simulation time. The second part applies this algorithm to several key physical devices: a flying qubit interferometer, a Josephson junction under bias, and a nanowire coupled to a superconductor, which exhibits Majorana states at its extremities. Here we shall briefly present the key results from each chapter.

Chapter 2: Introduction to Quantum Transport in the Time Domain

We start by introducing the field of quantum transport, including typical length scales, and look at the archetypal quantum device: a two-dimensional electron gas at the interface between two semiconductors. We then go on to look at recent experiments involving single electron sources, which highlight the need for a better understanding of the titular quantum transport in the time domain. We follow this with a brief review of the currently available numerical techniques: non-equilibrium Green's functions and wavefunction based approaches. This chapter does not contain any new results, but serves as an introduction to what follows.

Chapter 3: Numerical Algorithms for Time-Resolved Quantum Transport

In this chapter we present the improvements that have been made to the numerical algorithms with respect to the current state of the art. We start off by covering known material (calculation of stationary scattering wavefunctions) before presenting our

© Springer International Publishing AG 2017

J. Weston, *Numerical Methods for Time-Resolved Quantum Nanoelectronics*,
Springer Theses, DOI 10.1007/978-3-319-63691-7_1

approach to time-resolved transport. This consists of starting with initial scattering states of the system and then evolving them in time using the time-dependent
Schrödinger equation; observables are then calculated by integrating over the contributions from these time-evolved scattering states. In the end the differential equations
to solve are

$$i\frac{\partial}{\partial t}\bar{\psi}_{\alpha E}(t) = [\mathbf{H}(t) - E]\bar{\psi}_{\alpha E}(t) + \underbrace{\mathbf{W}(t)\psi_{\alpha E}^{st}}_{\text{source term}} - \underbrace{i\Sigma\bar{\psi}_{\alpha E}(t)}_{\text{sink term}} \tag{1.1}$$

$$\psi_{\alpha E}(t) = \left[\bar{\psi}_{\alpha E}(t) + \psi_{\alpha E}^{st}\right]e^{-iEt},$$

where $\psi_{\alpha E}(t)$ is the wavefunction in which we are interested, $\psi_{\alpha E}^{st}$ is the initial
scattering state incoming in channel α and energy E, $\mathbf{H}(t)$ is the Hamiltonian matrix,
$\mathbf{W}(t) = \mathbf{H}(t) - \mathbf{H}(0)$, and Σ is a diagonal matrix that is non-zero in a finite number of
cells of the leads that are attached to the central scattering region. The key difference
from previous approaches is the use of a *complex absorbing potential*, $-i\Sigma$, to handle
the boundary conditions at the system-lead interface. This is what gives our algorithm
linear scaling with system volume and simulation time. The characteristic "source"
and "sink" terms in the Schrödinger equation lead us to dub this the "source-sink"
method.

In addition we perform a detailed analysis of the effect of the complex potential,
including an analytical calculation of the reflection amplitude beyond the WKB
approximation. We find that the reflection from the complex potential satisfies

$$r_\Sigma(E) = ie^{2ikL}e^{-Ak/E} +$$

$$\frac{1}{4iEL}\int_0^\infty \Sigma'(u)\exp\left\{2ikLu - \frac{k}{E}\int_0^u \Sigma(v)dv\right\}du + \mathcal{O}\left(\left(\frac{1}{kL}\right)^2\right), \tag{1.2}$$

where $\Sigma(x)$ is the absorbing potential as a function of distance into the absorbing
region, L is the length of the absorbing region, and k is the wavevector corresponding
to energy E.

Finally, we discuss how the observables can be calculated by integrating the contributions from the wavefunctions, weighted by the appropriate Fermi-Dirac distribution. We propose to *integrate in momentum space*, as opposed to the energy-space
integrations used by competing methods, to avoid singularities arising from new
modes opening in the leads of the system. We provide two examples that illustrate
that the momentum-space integration requires drastically fewer points (and hence
fewer wavefunctions to evolve) than the corresponding energy-space integration.

Chapter 4: Software Design

In this chapter we discuss the requirements for a robust software tool that implements
the algorithms discussed in the preceding chapter. We start by motivating the need
for solid abstractions of the fundamental mathematical objects, as opposed to an
"all-singing all-dancing" monolithic code. We then take the example of the KWANT

package and illustrate—using a simple toy example—how it implements this philosophy for time-*independent* transport. In particular we put emphasis on how KWANT enables one to express a problem to solve *in terms of the mathematical structure*, instead of in terms of its low-level representation to the computer.

We move on to discuss the extra pieces that would have to be implemented on top of KWANT in order to be able to handle time-dependent problems, we dub these "extra pieces" TKWANT, for "time-dependent KWANT". We finish by showing a gallery of examples where *the current implementation of* TKWANT *has been used*, outside of the applications studied in this thesis: calculating time-resolved shot noise; stopping electrons in the quantum Hall regime; a universal transient regime for voltage pulses applied to interferometers; and simulating Floquet topological insulators. This includes work done outside of the research group of the author, which indicates that—despite its flaws—the current implementation is nevertheless *providing value* to research projects.

Chapter 5: Split Wire Flying Qubit

We now move to the second half of the thesis, which is concerned with specific applications of the aforementioned algorithms and software tools. The first application is to a split-wire setup implemented in a two-dimensional electron gas, which consists of two quasi one-dimensional regions separated by a controllable tunnelling barrier. This device has been proposed as an implementation of a "flying qubit", where the state of an electron is modified as it is moved around the quantum circuit. This device is currently being implemented experimentally in the group of Christopher Buäerle at the Néel Institute in Grenoble.

We start by treating the problem in the absence of any time dependence, using a scattering approach. This allows us to appreciate that the system acts as an *interferometer* with the symmetric and antisymmetric states in the split wire providing the two alternative paths through the system. We follow by applying a *pulse of bias voltage* to the split wire. We see that the number of charges recovered on the other side of the device *oscillates* with the number of charges sent by the voltage pulse, as shown in Fig. 1.1. Experimentally this would correspond to a measurement of the average current when the voltage pulse is repeated in time. This effect is interpreted within the paradigm of *dynamical control of interference* that was recently explored in a number of publications [1, 2]. The expressions in Ref. [1] for the number of particles transmitted in a two-path Mach-Zehnder interferometer are applied to the present case:

$$
\begin{aligned}
n_\uparrow &= \frac{\bar{n}}{2}\left[1 + \frac{1}{\pi}\sin(\pi\bar{n})\cos\left(\pi\bar{n} + \frac{\Delta k_0}{2}\tilde{L}\right)\right] \\
n_\downarrow &= \frac{\bar{n}}{2}\left[1 - \frac{1}{\pi}\sin(\pi\bar{n})\cos\left(\pi\bar{n} + \frac{\Delta k_0}{2}\tilde{L}\right)\right],
\end{aligned}
\tag{1.3}
$$

where $\bar{n}$ is the number of particles injected by the voltage pulse, Δk_0 is the difference between the wavevectors of the symmetric and antisymmetric wavefunctions at the Fermi level, and $\tilde{L}$ is the length over which the wires are coupled.

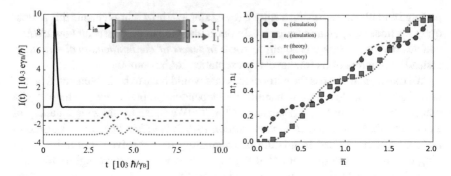

Fig. 1.1 Charge transport after application of a voltage pulse on one lead of the split wire setup (shown as an inset in subfigure (**a**)). **a** Output currents (*dashed* and *dotted lines*) and input current (*solid line*) in the split wire as a function of time. *Inset* sketch of the simulated setup. **b** Number of transmitted particles on the right in lead ↑ ($n_\uparrow$) and lead ↓ ($n_\downarrow$) as a function of the injected number of particles ($\bar{n}$). Symbols: time resolved simulation, dashed/dotted lines: analytical result

Chapter 6: **Time-resolved Dynamics of Josephson Junctions**

A Josephson junction with a voltage bias applied between the superconducting contacts is an inherently time-dependent system, which is illustrated by the appearance of the a.c. Josephson effect. This chapter deals with the dynamics of Josephson junctions with a static or time-varying bias voltage applied. Because of the large separation of energy scales required to study such a system in an experimentally relevant regime (the superconducting gap must be very small compared with the Fermi energy), the difference in time scales will be correspondingly large. This requirement to simulate to very long times (compared to the smallest time scale of the problem) is perfect for the source-sink algorithm due to the linear scaling of the latter.

After an introduction to the relevant parts of the theory of conventional superconductivity (Bogoliubov-de Gennes equation and Andreev reflection), we start by studying the multiple Andreev reflection (MAR) processes responsible for the finite current at voltages smaller than the superconducting gap. Despite the fact that we are using a time-resolved approach to study an essentially periodic problem, we nevertheless find *quantitative agreement* with theoretical results obtained using Floquet theory.

We then use the power of the time-resolved approach to study trains of voltage pulses propagating in long Josephson junctions. We see that a *periodic current*, is generated at the output even when just a single voltage pulse is applied, as can be seen in Fig. 1.2. The voltage pulse generates an excitation in the junction that becomes trapped, as the pulse is brief enough that the voltage is once again zero by the time the excitation traverses the junction and returns (after an Andreev reflection) to the contact where the pulse was applied.

We finally turn to short junctions, where the time of flight across the junction is much shorter than the duration of the voltage pulse, and see that we can still obtain a periodic current after a single voltage pulse. The pulse creates an excitation in a

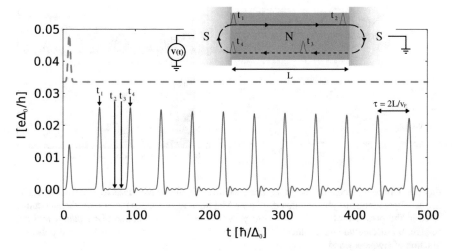

Fig. 1.2 Current (*solid line*) and voltage (*dashed line*, offset for clarity) at the left superconducting-normal contact as a function of time. *Inset* propagation of the charge pulse through the junction at different times (t_1, t_2, t_3, t_4) and the corresponding times indicated on the main plot

superposition of the pair of Andreev bound states in the junction, at energies E and $-E$, which gives rise to a current oscillating at frequency $2E/\hbar$.

Chapter 7: **Manipulating Andreev and Majorana Resonances in Nanowires**

The final application melds the concepts of interferometry introduced in Chap. 5 with those of superconductivity and Andreev bound states introduced in Chap. 6. We study a system consisting of a nanowire coupled to a superconductor, which have recently become a hot topic due to the presence of a Majorana state, of which a signature is a peak in the conductance at zero bias.

We start by treating the system in absence of the Majorana state, and show that by applying a *train of voltage pulses* we can manipulate the peaks in the differential conductance that are present at voltage below the superconducting gap due to the presence of Andreev resonances. In particular we can shift the resonances to different voltages when applying trains of different frequency.

Next we add further ingredients to the model (Rashba coupling and Zeeman coupling), and work in a parameter regime where spin-momentum locking is present, which gives rise to a Majorana state and a characteristic peak in the differential conductance at zero bias. We show that the same technique using a train of voltage pulses can be used to manipulate the Majorana resonance in the same way. We explore the effect of this train of pulses when the pulse amplitude and frequency are changed, and even use this to perform "spectroscopy" of the Majorana state, as illustrated in Fig. 1.3. This reveals distinct signatures for the resonant Andreev reflection mechanism that gives rise to the Majorana state.

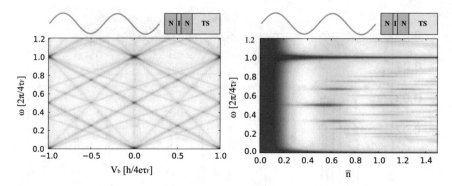

Fig. 1.3 Differential conductance in the presence of a train of voltage pulses for a normal-insulator-normal-superconductor junction that displays a Majorana resonance in d.c. **a** Differential conductance in the presence of a sinusoidal voltage with $\bar{n} = 0.5$ as a function of frequency and bias voltage. **b** Differential conductance at zero bias in the presence of a sinusoidal voltage pulse as a function of frequency and $\bar{n}$

This could be used as an *extra experimental probe* to show that a zero-bias conductance peak originates from resonant Andreev reflection, which could provide evidence of its Majorana character.

References

1. B. Gaury, X. Waintal, Dynamical control of interference using voltage pulses in the quantum regime. Nat. Commun. **5**, 3844 (2014)
2. B. Gaury, J. Weston, X. Waintal, The a.c. Josephson effect without uperconductivity, Nat. Commun. **6**, 6524 (2015)

Part I
Numerical Algorithms and Software for Time-Resolved QuantumTransport

Chapter 2
Introduction to Quantum Transport in the Time Domain

The study of quantum transport is the study of the flow of electrons through small electronic circuits, typically of a few microns (μm) in extent and cooled to cryogenic temperatures <1 K. At such small length scales, and at such low temperatures, the electrons behave according to the laws of quantum mechanics, which gives rise to behaviour that is *qualitatively different* compared to the classical behaviour, due to the fact that the electrons now behave as waves. The field of quantum transport is now entering a new era as it becomes possible to modify these circuit devices on shorter and shorter time scales. In practice this could involve applying a quickly-changing bias voltage across the device or rapidly charging and discharging a nearby capacitor. Operating on ever shorter timescales allows us to access qualitatively different regimes of operation for these devices, where we can start to probe the *internal dynamics* far beyond the adiabatic limit.

In this thesis we are concerned with the theoretical and numerical techniques required for treating this so-called "time-resolved quantum transport". This chapter contains a general introduction to the field of quantum transport, discussing the relevant length and time scales, before moving on to the recent experimental progress that serves as a motivation for studying the emerging sub-field of time-resolved transport. Finally, we discuss existing theoretical and numerical techniques for treating problems in this field.

2.1 Mesoscopic Quantum Electronics

Let us begin by getting a general feel for the sort of devices and length scales with which we will be concerned in this thesis. In general we will be studying the *coherent* transport of electrons, that is, where their quantum-mechanical wave-like nature is exhibited; this is also referred to as *quantum* transport. This already puts an upper bound on the size of circuit that we wish to consider. If we want to be able to observe

© Springer International Publishing AG 2017
9
J. Weston, *Numerical Methods for Time-Resolved Quantum Nanoelectronics*,
Springer Theses, DOI 10.1007/978-3-319-63691-7_2

quantum effects (notably interference), the phase of the electronic wavefunction must remain well-defined across the system. At distances greater than L_φ—the *coherence length*—the electronic wavefunction will lose its well-defined phase; the characteristic quantum interference will tend to be washed out. The physical origin of the finite coherence length is related to interactions of the electrons with other degrees of freedom in the material (e.g. lattice vibrations, impurities with some internal degrees of freedom, or electron-electron interactions) [1, 2]. Naturally, then, the coherence length will strongly depend on temperature; we will normally have to descend to cryogenic temperatures (< 1 K) in order to "freeze out" the non-electronic degrees of freedom that will give rise to decoherence. At these temperatures coherence lengths of the order of tens of μm have been measured experimentally in certain semiconductor heterostructures [3]. In this thesis we will not be concerned with the process of decoherence per se, however we will always have to remember to place ourselves in an appropriate parameter regime (with respect to system size and temperature) so that decoherence is not an issue.

We will also impose a lower-bound on the length scales of interest to us; we do not want to describe details on the scale of single atoms. While devices such as molecular junctions—where a molecule is suspended between large metallic contacts—can, in principle, be described by the techniques that we will present [4], this is not our domain of interest. We will mostly be interested in cases where the electrons in a material "see" the underlying ionic lattice as a continuum, and the specific material properties enter only in the effective mass of the electrons [1]. This is valid when the Fermi wavelength of the electrons is large compared to the inter-atomic distance. This range of distances, of the order of a few μm but larger than atomic distances, is referred to as the "mesoscopic" scale.

Another key feature of electronics at the mesoscopic scale is that devices are usually constructed so that the electronic motion is restricted in one or more spatial dimensions. For the electrons, the circuit is effectively two or one dimensional, even though the actual device obviously exists in three dimensions. The archetypal mesoscopic system is the two-dimensional electron gas (2DEG) that forms at the interface between layers of aluminium gallium arsenide and gallium arsenide; a sketch is shown in Fig. 2.1a. Figure 2.1a also shows a simplified sketch of the valence/conductance bands at the interface of such a heterostructure; we see that the charge transfer that equalises the Fermi level on either side of the interface induces an electric field that confines electrons close to the interface. The electronic confinement along the z direction leads to quantization of the z component of the quasi momentum p_z, although the electrons are still quasi-free in the $x - y$ plane parallel to the interface. This quantization effectively "freezes out" the z degree of freedom, as long as any perturbations made to the 2DEG are small compared to the energy required to transition to a state with different p_z. Figure 2.1b shows a scanning electron microscope image of a mesoscopic circuit constructed from such a heterostructure. The 2DEG is in a plane parallel to the image, embedded $\sim$100 nm below the surface. The lighter grey rectangles are made of metal deposited on top of the heterostructure, and are referred to as "gates". As the gates are separated from the 2DEG by a layer of semiconductor (which is insulating) no electrons flow between

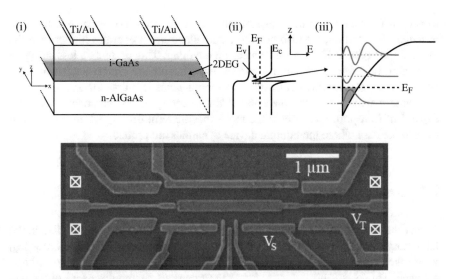

Fig. 2.1 Illustrations of a 2DEG; a conceptual picture, and the experimental reality. **a** (i) Sketch of an AlGaAs/GaAs heterostructure with a 2DEG at the interface and metallic Ti/Au gates deposited on the surface. (ii) Sketch of the conduction (E_c) and valence (E_v) bands in the vicinity of the interface. The "+" symbols show the positively charged donors and the 2DEG is indicated in grey. (iii) A sketch of the quantised modes in the z direction; in this example only the lowest mode is populated as the others are above the Fermi energy (E_F). **b** Scanning electron microscope image of a flying qubit interferometer in an AlGaAs/GaAs 2DEG. The 2DEG is in a plane parallel to the page and roughly 100 nm below it. The *lighter grey regions* are metallic gates deposited on the surface of the heterostructure. Reprinted with permission from Ref. [6], copyright 2015 by the American Physical Society

the gates and the 2DEG. If a voltage is applied to a gate, however, the electrons in the 2DEG will feel the electric field produced; this can be used to confine the electrons within subregions of the 2DEG. The white squares in Fig. 2.1b indicate where electrons will be able to flow in/out of the 2DEG through ohmic contacts [5] into metallic *leads* (we will also refer to these as *electrodes* or *contacts*). These leads interface the quantum circuit with the macroscopic world, which consists of the measurement apparatus, voltage sources, radio-frequency transmission lines etc.

In this thesis we will be developing and applying numerical techniques to simulate the behaviour of these sorts of mesoscopic devices when their controlling parameters (such as the gate or bias voltages discussed above) are modified quickly enough to probe the internal dynamics of the device. Concretely this means varying the control parameters quickly compared to the time it takes an electron to traverse the device. In the 2DEGs discussed above the electrons at the Fermi level typically travel at speeds of 10^4–10^5 m s^{-1} [7], which means that the control parameters need to vary at frequencies in the range of tens of GHz, when the device is a few μm in length. In addition, we also need to excite electrons at energies higher than the thermal background if we hope to measure anything. This presents a less stringent constraint, however, as quantum transport experiments are typically carried out at temperatures $\ll 1$ K, which corresponds to frequencies less than $20\,\mathrm{GH}_Z$. We refer to quantum transport in the presence of time-varying device parameters as time-*dependent* transport, and reserve the more specific term time-*resolved* transport to refer to the case where the internal device dynamics are probed.

2.2   Experiments in the Time Domain

One of the first examples of time-*dependent* quantum transport being studied in the laboratory was the measurement of photo-assisted tunnelling by Tien and Gordon [8], where the presence of an a.c. bias voltage affects the d.c. current flowing through a device. This was followed, at around the same time, by the discovery of the a.c. Josephson effect [9, 10], where a d.c. bias voltage causes an a.c. output current in a superconducting junction. Over the years several other novel effects at finite frequency were discovered, such as charge pumping [11, 12] (where a purely a.c. voltage with no d.c. component can induce a d.c. current).

The recent move towards time-*resolved* transport has been motivated by the desire to build coherent sources of single electrons. To see why these two ideas are linked, let's consider the application of a finite, static bias to an electrode of a quantum circuit. This can be seen as producing a continuous stream of electrons that flow from the biased electrode to be collected by the other (grounded) electrodes. If we now apply the bias only during a finite time interval, we will clearly only transfer a finite number of charges. As we reduce the time over which we apply the bias we will eventually arrive at the point where the bias "pulse" is so brief that only a single charge is transferred. We refer to such weak/brief bias pulses as being "in the quantum regime" when they only excite one or a few charges, i.e.

$$n = \frac{e}{h} \int V(t)\mathrm{d}t, \tag{2.1}$$

where n is a small integer. Although in practice the generation of coherent single electrons is more complicated than this naïve picture, it nevertheless motivates why a time-*resolved* description will be necessary.

The first single-electron sources were realised using a different paradigm to the one outlined above. Instead of applying a bias voltage to inject electrons from an electrode into the quantum device, gates were used confine electrons in a region of a 2DEG with size comparable to the Fermi wavelength [13–15]. As a result the electrons in this so-called "quantum dot" have their energy quantised; A gate applied

(a)   **(b)**

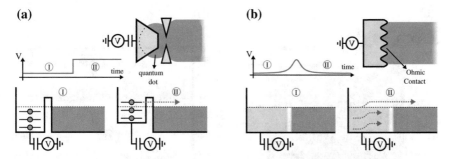

Fig. 2.2 Two techniques for producing coherent single-electron excitations in quantum circuits, that were both recently realised experimentally. **a** Single-electron source using a quantum dot in a 2DEG. The schematic (*top-right*) shows how the gates (*light grey*) constrict the 2DEG (*dark grey*) to form a quantum dot with discrete energy levels. The voltage V applied to the top-gate shifts the levels of the dot to bring a single electron above the Fermi energy in the rest of the 2DEG; the electron tunnels through the barrier and escapes. **b** Single-electron source using a lorentzian pulse applied to an ohmic contact. The schematic (*top-right*) shows the ohmic contact (*light grey*) via which charges can be *injected* from the lead into the attached to the 2DEG (*dark grey*). The bias voltage pulse V excites the Fermi sea of the lead; the specific form of the pulse ensures that when $(e/h) \int V(t) \mathrm{d}t = 1$ the net result is that only a *single electron* is excited, and the Fermi sea remains undisturbed (see main text)

to the top of the confined region is used to shift the energy levels of the underlying quantum dot so that a single electron is brought above the Fermi energy of the surrounding 2DEG. The electron can then tunnel through the confining potential and propagate into the rest of the 2DEG, as illustrated in Fig. 2.2a. This setup can produce single electrons with well-defined energy, but poorly defined release time (due to the Heisenberg uncertainty relation $\Delta E \Delta t > \hbar/2$). Such single-electron sources were used to probe the electronic wave-particle duality in a Hanbury-Brown-Twiss (HBT) setup [16]. Additionally the fermionic nature of electrons was visualised through anti-bunching behaviour in a Hong-Ou-Mandel (HOM) setup [17].

This last experiment highlights the importance of the *fermionic* nature of the electrons when treating such mesoscopic devices. As this point will be important for our discussion of theoretical methods for time-resolved transport, we shall look at the experiment of Ref. [17] a bit more closely. Figure 2.3a shows an annotated electron microscope image of the experimental setup. A 2DEG (dark grey) is attached to several electrodes (white boxes), and gates (light grey fingers) constrict the 2DEG at the location marked "beamsplitter", which will cause the electronic wavefunction to be partially reflected. The gates marked "source 1" and "source 2" are cover the quantum dots, which host electrons that can be excited into the surrounding 2DEG by raising the gate voltage (i.e. the single-electron sources originally realised in Ref. [13]). A perpendicular magnetic field is applied to the device, which causes the electrons to propagate in unidirectional edge channels in the 2DEG[1] (shown as paths with arrows in Fig. 2.3a). The idea of the experiment is to send voltage pulses onto the

---

[1]This is the quantum Hall effect, which will be briefly discussed in Sect. 4.3.2.

(a)   **(b)**

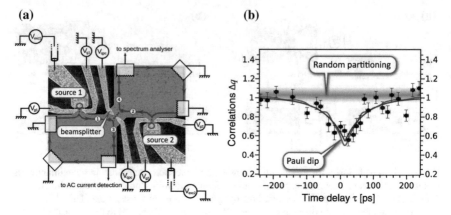

Fig. 2.3 Experimental results from Ref. [17], showing how the fluctuations in the electric current are affected when two identical electron wavepackets are incident on opposite sides of an electronic beam splitter. When the delay between the arrival of the wavepackets is small, there is a dip in the current fluctuations due to the Pauli principle. Both subfigures are from Ref. [17] and reprinted with permission from AAAS. **a** Annotated electron microscope image of the experimental setup. Metallic gates (*light grey*) are capacitively coupled to an embedded 2DEG (*dark grey*). Ohmic contacts (*white boxes*) measure the output current. **b** Excess noise in the number of transmitted particles as a function of the delay between the emission of single electrons from source 1 and source 2

gates "source 1" and "source 2" with a slight delay relative to one another. This will mean that two electrons will be excited above the Fermi sea, one at source 1 and the other at source 2, which will begin propagating towards the beamsplitter (along paths 1 and 2 in Fig. 2.3a). As the voltage pulse is sent to source 2 with a delay with respect to the voltage pulse sent to source 1, the electrons from the two sources will arrive at the beamsplitter with a corresponding delay. In the case where the electrons arrive at the beam splitter at the same time they must exit along *different* paths (3 or 4); if they exit along the same path they would be in identical states, which is disallowed due to the Pauli principle. There are two possibilities: both electrons are transmitted, or both particles are reflected, in either case each of the contacts on the paths 3 and 4 will receive *exactly 1 electron*. If the electrons arrive slightly delayed then it is possible that they both exit the beamsplitter along the same path, as they will not be perfectly overlapping (and hence not in the same state) in this case. Unfortunately it is not yet experimentally possible to have one-shot detection of ballistically propagating single electrons in condensed matter. Instead, experimentalists will typically generate *many* single-electron excitations one after the other and then measure the *average* current, as well as its noise properties.[2] In the HOM setup the Pauli principle should then manifest itself in a reduction of the current noise when there is no delay between the arrival of the electrons; this is shown in Fig. 2.3b.

---

[2]The time delay between the subsequent single-electron excitations should, therefore, be much greater than the time it takes for an electron to traverse the device.

Later, another method of producing coherent single-electron excitations was demonstrated [18, 19]. Here, instead of initially confining electrons in a quantum dot, a Lorentzian-shaped voltage pulse applied to an electrode coupled to the 2DEG via an ohmic contact *injects* a single-electron excitation from the electrode into the 2DEG. Figure 2.2b shows an illustration of this approach, which can be compared to the quantum dot approach discussed previously. These experiments were motivated by the seminal work of Levitov [20, 21], who showed that the *shape* (or equivalently, the harmonic content) of the voltage pulse is of tantamount importance in the generation of coherent single-electron excitations. Applying an arbitrary voltage pulse will, in general, perturb the Fermi sea and produce excitations above the Fermi energy (electron-like excitations) as well as below (hole-like excitations), which will both propagate into the device. Levitov showed that when a *Lorenzian* pulse is used only electron-like excitations above the Fermi energy will propagate into the device.[3] Such a Lorentzian pulse $V(t) = V_p/[(t - t_0)^2 + 1]$ can excite a *single electron*, above an undisturbed Fermi sea, when $(e/h) \int V(t)dt = 1$. Such excitations are referred to as *Levitons*. Owing to the continuum of energy states in the electrode (as opposed to the discrete levels of the quantum dot), the excitation is poorly resolved in energy but *well* resolved in time. We should emphasise that this is a *completely* different paradigm for generating single-charge excitations than the quantum dot approach of Ref. [13]. While the approach using quantum dots can be understood as tuning the levels in the dot to put a single level (and hence single electron) above the surrounding Fermi sea, the Leviton is instead a *collective excitation* of the Fermi sea itself. That such a collective excitation consists, in the end, of an unperturbed Fermi sea with unentangled, purely electron-like excitations on top is far from obvious [21].

After their experimental discovery, Levitons were then used in HBT and HOM setups [19] analogous to those of Refs. [16, 17]. Figure 2.4a shows an artist's impression of the HOM experimental setup, and the results of current noise measurements are shown in Fig. 2.4b. We clearly see that the noise drops to zero when the two Levitons arrive at the beamsplitter with no time delay, which is due to the Pauli principle, as discussed previously. Even more recent experiments used shot noise measurements to directly reconstruct the temporal structure of the Leviton wavefunction [22].

This collection of experiments adds yet more techniques to the toolbox of the emerging field of "electronic quantum optics" [23], where quantum optics experiments are performed with electrons. Such devices could have wider applications in the field of quantum computing [24–26]. What is clear is that theoretical and numerical techniques are required to explore the inherently time-resolved nature of these experiments, as well as to propose new ones.

---

[3]The hole-like excitations, required to maintain charge balance, move in the opposite direction and do not enter the device proper.

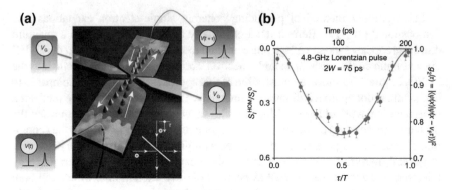

Fig. 2.4 **a** Artist's impression of a Hong-Ou-Mandel setup in a 2DEG (*dark grey*) connected via ohmic contacts (*light grey rectangles*) to metallic leads. A voltage V_g is applied to metallic gates (*light grey fingers*) to constrict the 2DEG and induce electronic backscattering. Voltage pulses $V(t)$ are applied to the leads, with a time delay τ between the pulses on the upper/lower contact. **b** Current noise measured in a single contact, as a function of the delay τ between the voltage pulses applied on each of the ohmic contacts. The noise falls to zero when there is no delay between the pulses; the fermionic nature of the single electron excitations means that each contact receives exactly one unit of charge each time a pair of electrons are injected. Figures reprinted from Ref. [19] with permission

2.3 Theoretical Description of Quantum Circuits

2.3.1 A General Model for Quantum Circuits

So now we have a bit of an idea about why it might be interesting to study time-resolved quantum transport. The next question is *how* can we study such a problem? The general class of systems that we wish to study in this thesis will consist of a number of quasi one-dimensional leads (collectively referred to as L) connected to a central device S. The basic pieces of information that we need to study such a setup are the Hamiltonian of the leads, $\hat{H}^L$, the Hamiltonian of the central device, $\hat{H}^S(t)$, and the lead-device coupling, $\hat{H}^T(t)$. Even though there may be multiple physically separate leads, we regroup all the lead degrees of freedom into a single $\hat{H}^L$. Given that we will be treating transport *through* the device, where the number of charges in the device is not fixed, it will be easiest to express the Hamiltonian using the language of second quantisation. The fundamental objects of this language are "creation" and "annihilation" operators $\hat{c}_i^\dagger$ and $\hat{c}_j$ that act on the full space of many-particle states. We aim to write expressions for observable quantities in terms of these operators, rather than in terms of the many-body wavefunctions. Although this is formally equivalent to using wavefunctions, the expressions involving the creation/annihilation operators are considerably more compact. Reference [27] contains a good introduction to this topic.

In this thesis we will not be concerned with modelling electron-electron inter-actions. This means that all the Hamiltonians that we will consider only contain terms that are bilinear in creation/annihilation operators (these are also referred to as "quadratic" Hamiltonians). In addition, as we will be wanting to simulate such systems on a computer, the Hamiltonians will only involve a *discrete* set of degrees of freedom (although we will be able to treat *infinite* sub-systems by exploiting translational symmetry); these are referred to as *tight-binding* models. Putting these ingredients together we can write down the most general form of Hamiltonian that we wish to treat:

$$\hat{H}(t) = \underbrace{\sum_{i,j\in S} H_{ij}(t)\hat{c}_i^\dagger \hat{c}_j}_{\hat{H}^S(t)} + \underbrace{\sum_{i\in S, j\in L} H_{ij}(t)\hat{c}_i^\dagger \hat{c}_j + h.c.}_{\hat{H}^T(t)} + \underbrace{\sum_{i,j\in L} H_{ij}\hat{c}_i^\dagger \hat{c}_j}_{\hat{H}^L}. \qquad (2.2)$$

The $\hat{c}_i^\dagger$ ($\hat{c}_j$) are operators that create (destroy) electrons in a single-particle state enumerated by the index i (j), which we refer to as a *site*. The site index may label position as well as other degrees of freedom such as spin or orbital angular momentum, although in specific cases we shall often make an explicit distinction between spatial and internal degrees of freedom. The $H_{ij}(t)$ are time-dependent complex numbers that we collectively refer to as the matrix

$$\mathbf{H}(t) = \begin{pmatrix} \mathbf{H}^S(t) & \mathbf{H}^T(t) \\ [\mathbf{H}^T(t)]^\dagger & \mathbf{H}^L \end{pmatrix} \qquad (2.3)$$

where the sub-matrices are the device ($\mathbf{H}^S$), lead ($\mathbf{H}^L$), and coupling terms ($\mathbf{H}^T$).

Given that the Hamiltonian is fully characterised by the matrix $\mathbf{H}(t)$, which is just the Hamiltonian in first quantisation, naïvely one may think that we just need to use the time-dependent Schrödinger equation on some wavepacket initial state and call it a day. The situation is, however, a little more complicated than this. The complica-tions arise due to two aspects peculiar to *open, fermionic* systems. Firstly the *open* condition means that we treat the electrodes as being infinite in extent (though, for simplicity, we shall always treat them as being periodic). This has profound conse-quences on a mathematical level, as the spectrum of the Hamiltonian will now have a *continuous* part (that will mostly dominate the transport properties), in addition to a discrete part. Secondly, there is a filled Fermi sea already present in the system that—as we saw previously in the Hong-Ou-Mandel experiment—is crucial to obtaining the correct physics. In a system with time-dependent perturbations electrons may be excited to different energies, however the Fermi sea prevents certain transitions (to already filled states) from being possible. It is not immediately obvious how this con-dition can be satisfied just by solving the single-particle time-dependent Schrödinger equation. We shall see that it is the presence of the filled Fermi sea that will give us the correct initial conditions for the problem in terms of the macroscopic (and directly experimentally controllable) parameters of the system, rather than resorting to microscopic details in the form of electronic wave packets.

2.3.2 Non-equilibrium Green's Function Techniques

Historically the first class of techniques to deal with the these two issues were the non-equilibrium Green's function (NEGF) techniques. The earliest numerical simulations of time-resolved transport were based on a seminal article by Caroli et al. [28], which applied the Keldysh formalism [29] to a strictly one-dimensional (single mode) model. This technique was applied in Ref. [30] to study resonant tunneling through a device consisting of a single site. A more general formulation for generic device geometries was later proposed by Wingreen et al. [31] and Jauho et al. [32], following their own extension of the stationary non-equilibrium formalism [33], which was itself based on Ref. [28]. The formalism described in these papers is at the foundation of the non-equilibrium Green's function techniques used today.

In the NEGF approach the fundamental objects are *correlators* (called Green's functions) between the electron creation/destruction operators introduced previously. Although a whole zoo of such correlators exist, two of the most important ones are the so-called *retarded* ($\mathcal{G}_{ij}^R(t, t')$) and *lesser* ($\mathcal{G}_{ij}^<(t, t')$) Green's functions:

$$\mathcal{G}_{ij}^R(t, t') = -i\Theta(t - t')\langle\{\hat{c}_j^\dagger(t'), \hat{c}_i(t)\}\rangle \tag{2.4}$$

$$\mathcal{G}_{ij}^<(t, t') = i\langle\hat{c}_j^\dagger(t')\hat{c}_i(t)\rangle \tag{2.5}$$

where $\Theta(t - t')$ is the Heaviside function, $\{\cdot, \cdot\}$ denotes an anticommutator, the $\hat{c}_j^\dagger(t)$ and $\hat{c}_i(t)$ are creation/destruction operators in the Heisenberg picture [27]:

$$\hat{c}_j^\dagger(t) = \hat{U}(t)\,\hat{c}_j^\dagger\,\hat{U}^\dagger(t) \tag{2.6}$$

$$\hat{c}_i(t) = \hat{U}(t)\,\hat{c}_i\,\hat{U}^\dagger(t), \tag{2.7}$$

with $\hat{U}(t)$ the evolution operator satisfying $i\partial_t\hat{U}(t) = \hat{H}(t)\hat{U}(t)$, and $\langle\cdot\rangle$ denotes a thermal average, i.e. $\langle\hat{A}(t)\rangle = \mathrm{Tr}\left[e^{-\hat{H}(0)/k_B T}\hat{A}(t)\right]$, where $\hat{A}(t)$ is a Heisenberg-picture operator, and k_B and T are the Boltzmann constant and the temperature.[4] It turns out that all the one-body observables can be calculated from $\mathcal{G}_{ij}^<(t, t')$: for example the electron density on site i is $\rho_i(t) = -i\mathcal{G}_{ii}^<(t, t)$, and the average current between sites i and j can be written

$$I_{ij}(t) = H_{ij}(t)\mathcal{G}_{ji}^<(t, t) - H_{ji}(t)\mathcal{G}_{ij}^<(t, t). \tag{2.8}$$

Given that we are only interested in evaluating quantities within the device region or currents flowing between the device and the electrodes, we only need elements of

---

[4]This expression can be made slightly more general by replacing the operator exponential by a generic density operator, but the thermal one presented in the main text is the only one we consider in this thesis.

$\mathcal{G}_{ij}^R(t, t')$ with both indices in the device region—$i, j \in S^5$—which we shall denote $G_{ij}^R(t, t')$ (similarly $G_{ij}^<(t, t')$ for the lesser Green's function). The equations of motion satisfied by $G_{ij}^R(t, t')$ (and its relation to $G_{ij}^<(t, t')$) can be written [27, 34]:

$$i \frac{\partial}{\partial t} \mathbf{G}^R(t, t') = \mathbf{H}^S(t) \mathbf{G}^R(t, t') + \int du \, \Sigma^R(t, u) \mathbf{G}^R(u, t'), \qquad (2.9)$$

$$\mathbf{G}^<(t, t') = \int \int du \, dv \, \mathbf{G}^R(t, u) \Sigma^<(u, v) \left[\mathbf{G}^R(t', v) \right]^\dagger. \qquad (2.10)$$

where $\mathbf{G}^R(t, t')$ and $\mathbf{G}^<(t, t')$ are matrices with elements $G_{ij}^R(t, t')$ and $G_{ij}^<(t, t')$ respectively. The quantities $\Sigma^R(t, t')$ and $\Sigma^<(t, t')$ are the so-called retarded and lesser *self-energies* that take into account the effect of the leads, and are defined by

$$\Sigma^{R(<)}(t, t') = \mathbf{H}^T(t) \mathbf{g}^{R(<)}(t, t') \left[\mathbf{H}^T(t) \right]^\dagger, \qquad (2.11)$$

where $\mathbf{g}^{R(<)}(t, t')$ is the retarded (lesser) Green's function for the leads *in isolation*, i.e. in the absence of coupling to the device region. Generally the problem to solve, therefore, takes the form of coupled integro-differential equations for several large, dense matrices.

There has been a great deal of effort over the years to design efficient strategies to integrate these equations of motion [35–39], including recursive techniques [40] and replacing the convolution-type integral with complex absorbing boundary conditions [41]. In addition the issues involved with properly including electron-electron interactions has been discussed [42–46]. Others have also derived semi-analytical expressions to calculate restricted parts of the full Green's function in specific physical situations [47–51]. An alternative but related approach introduced by Cini [52] does something a little different from the above-defined NEGF, in that one starts at $t = 0$ with the exact density matrix for the full problem and follows the system states as they are driven out of equilibrium. More recent work developed Green's function techniques within this framework [53, 54]. All these approaches are, however, fundamentally limited by the fact that the equations of motion involve dense matrices, whose number of elements scale as $\mathcal{O}(N^2)$, where N is the number of sites in the device region.

2.3.3 Wavefunction Techniques

In NEGF the Green's functions are the fundamental objects of the theory. An alternative consists in dealing directly with the many-body wavefunction of the system by

---

[5]We can always re-define what we consider "the device" in order to calculate currents flowing into the leads. The periodicity of the leads ensures that there will be no backscattering within the leads themselves.

calculating single-particle wavefunctions. In a non-interacting system the full many-body wavefunction is formed from a simple Slater determinant of single-particle wavefunctions, so it is at least reasonable that such an approach could be equivalent to NEGF. Nevertheless it was not clear until relatively recently [55] that there was a formal equivalence between the two approaches because of the aforementioned issues of dealing with infinite systems, as well as the Pauli principle. The use of wavefunction-based methods has recently gained popularity [56–60], with variations on the theme using a "stroboscopic" wavepacket basis [61, 62], or absorbing boundary conditions [63, 64].

In order to get a feel for how such an approach works in principle, it will be illustrative to look at a simplified example consisting of a *finite* system S at zero temperature. This will show explicitly how the Pauli principle is satisfied due to the fact that the unitary evolution of the single-particle states guarantees their mutual orthogonality at all times. We shall start with a similar model to that of Eq. (2.2), but without leads:

$$\hat{H}(t) = \sum_{i,j \in S} H_{ij}(t) \hat{c}_i^\dagger \hat{c}_j. \tag{2.12}$$

We shall assume that the time-dependence is only switched on for $t > 0$, so that we can diagonalise $\hat{H}(t \leq 0)$:

$$\hat{H}(t \leq 0) = \sum_\alpha E_\alpha \hat{d}_\alpha^\dagger \hat{d}_\alpha \tag{2.13}$$

where

$$\hat{d}_\alpha^\dagger = \sum_j [\varphi_\alpha]_j \, \hat{c}_j^\dagger \tag{2.14}$$

and the φ_α are column vectors of complex numbers (with jth element $[\varphi_\alpha]_j$) that satisfy time-independent Schrödinger equations $\mathbf{H}(t \leq 0)\varphi_\alpha = E_\alpha \phi_\alpha$. As we are at zero temperature, the full many-body state at $t \leq 0$ is just the state where all single-particle states below the Fermi energy E_F are filled:

$$|\Psi_0\rangle = \prod_{E_\alpha < E_F} \hat{d}_\alpha^\dagger |0\rangle, \tag{2.15}$$

where $|0\rangle$ is the vacuum state. The antisymmetry of $|\Psi_0\rangle$ under particle exchange is guaranteed by the anticommutation relations satisfied by the operators $\hat{d}_\alpha$:

$$\{\hat{d}_\alpha, \hat{d}_\beta\} = 0 \;,\; \{\hat{d}_\alpha, \hat{d}_\beta^\dagger\} = \delta_{\alpha\beta}, \tag{2.16}$$

where $\delta_{\alpha\beta}$ is the Kronecker delta. If we now look at times $t > 0$ the many-body state will evolve to

$$|\Psi(t)\rangle = \hat{U}(t) \prod_{E_\alpha < E_F} \hat{d}_\alpha^\dagger |0\rangle, \tag{2.17}$$

where $\hat{U}(t)$ is the evolution operator, which satisfies $i\partial_t \hat{U}(t) = \hat{H}(t)\hat{U}(t)$. As $\hat{U}(t)$ is unitary, we can sandwich factors of $\hat{U}^\dagger(t)\hat{U}(t)$ between the $\hat{a}_\alpha^\dagger$'s, and use the property $\hat{U}(t)|0\rangle = |0\rangle$ to write

$$|\Psi(t)\rangle = \prod_{E_\alpha < E_F} \hat{a}_\alpha^\dagger(t)|0\rangle, \qquad (2.18)$$

where $\hat{a}_\alpha^\dagger(t)$ is *defined* as

$$\hat{a}_\alpha^\dagger(t) \equiv \hat{U}(t)\, \hat{a}_\alpha^\dagger \, \hat{U}^\dagger(t). \qquad (2.19)$$

Clearly the $\hat{a}_\alpha^\dagger(t)$ satisfy the equal-time anticommutation relations

$$\{\hat{a}_\alpha(t), \hat{a}_\beta(t)\} = 0 \;,\;\; \{\hat{a}_\alpha(t), \hat{a}_\beta^\dagger(t)\} = \delta_{\alpha\beta}, \qquad (2.20)$$

which ensures that $|\Psi(t)\rangle$ is fully antisymmetric under particle exchange at any time t, and hence satisfies the Pauli principle at all times. We can also choose to write $\hat{a}_\alpha^\dagger(t)$ in terms of the original $\hat{c}_j^\dagger$'s

$$\hat{a}_\alpha^\dagger(t) = \sum_j [\psi_\alpha(t)]_j \; \hat{c}_j^\dagger, \qquad (2.21)$$

where $\psi_\alpha(t \leq 0) = \psi_\alpha$. Applying the operator $i\partial_t$ to Eqs. (2.19) and (2.21), and equating the right-hand sides, we get (after some algebra)

$$\sum_j i\left[\frac{\partial}{\partial t}\psi_\alpha(t)\right]_j \hat{c}_j^\dagger = \sum_j \sum_k \mathbf{H}_{jk}(t)\,[\psi_\alpha(t)]_k \; \hat{c}_j^\dagger \;, \quad \psi_\alpha(t \leq 0) = \psi_\alpha. \qquad (2.22)$$

As the $\hat{c}_j^\dagger$ are creation operators for *mutually orthogonal* single-particle states, we can equate the coefficients for each term j of the above sum, from which we see that $\psi_\alpha(t)$ *satisfies a time-dependent Schrödinger equation* $i\partial_t\psi_\alpha(t) = \mathbf{H}(t)\psi_\alpha(t)$. This means that in order to calculate the full many-body evolution we actually only need to solve n Schrödinger equations for all the *single-particle* states with initial energy less than the Fermi energy. Expectation values of one-body observables also take a simple form. For example, the average number of particles on site i, defined as $\langle\Psi(t)|\hat{c}_i^\dagger\hat{c}_i|\Psi(t)\rangle$ (for the zero-temperature case) can be written (after some algebra)

$$\langle\Psi(t)|\hat{c}_i^\dagger\hat{c}_i|\Psi(t)\rangle = \sum_{E_\alpha < E_F} |\psi_\alpha(t)|_i^2. \qquad (2.23)$$

The expectation value of one-body operators can be written as the sum of the expectation values of the associated operators in first-quantisation, evaluated on the states with energy less than the Fermi energy. When the temperature is instead finite, the result is similar, except that the terms in the sum Eq. (2.23) are weighted by the appropriate (Fermi-Dirac) occupation factor $f(E_\alpha)$.

If we also allow the system to now contain an *infinite* number of degrees of freedom (i.e. we attach leads), the discrete energies E_α will form a continuum energy *band* and the sum Eq. (2.23) will be replaced by an integral. This last step of reasoning is not presented in a particularly rigorous way here; a proper derivation for the case of infinite systems (which also contains the case when the system starts in an out-of-equilibrium steady state) can be found in Ref. [55].

The result (in the absence of true bound states in the system) is that the thermal average of an observable $\hat{A} = \Sigma_{ij} A_{ij} \hat{c}_i^\dagger \hat{c}_j$ at time t can be calculated as

$$\langle \hat{A}(t) \rangle = \sum_\alpha \int_{B_\alpha} \frac{dE}{2\pi} \, f_\alpha(E) \psi_{\alpha E}(t)^\dagger \mathbf{A} \, \psi_{\alpha E}(t). \tag{2.24}$$

The $\psi_{\alpha E}(t)$ are time-evolved single-particle states that, at $t = 0$, were so-called *scattering states* of the system. Scattering states extend infinitely far into the leads of the system, and have the particular characteristic that their wavefunction in all of the leads is propagating *away* from the central region, except in a single lead where there is an *incoming* component (characterised by a "mode index" α). A more precise definition of scattering states is given in Chap. 3, as well as details of how to calculate them for the initial system.

The energy integral runs over the energy band B_α of the mode α: $B_\alpha = \big(\inf E_\alpha(k),$ sup $E_\alpha(k) \big)$ where $E_\alpha(k)$ is the dispersion relation of mode α and k runs over the Brillouin zone. $f_\alpha(E)$ is the Fermi-Dirac function associated with the lead to which the mode α belongs (different leads may have different temperatures and/or Fermi energies). We can also write the retarded and lesser Green's functions in terms of the wavefunctions [55]:

$$G_{ij}^R(t, t') = -i\Theta(t - t') \sum_\alpha \int_{B_\alpha} \frac{dE}{2\pi} \, [\psi_{\alpha E}(t)]_i \, \left[\psi_{\alpha E}^\dagger(t') \right]_j \tag{2.25}$$

$$G_{ij}^<(t, t') = i \sum_\alpha \int_{B_\alpha} \frac{dE}{2\pi} \, f_\alpha(E) \, [\psi_{\alpha E}(t)]_i \, \left[\psi_{\alpha E}^\dagger(t') \right]_j , \tag{2.26}$$

which forms the link between the wavefunction and Green's function based approaches.

In the next chapter we will build on the specific wavefunction method introduced in Ref. [55]. We choose to develop an entirely wavefunction based approach due to the superior scaling properties with respect to the system size and simulation time compared to direct Green's function approaches. This essentially comes down to the fact that the single-particle wavefunctions are *vectors* with N elements, as opposed to the dense matrices (N^2 elements) of the Green's function approach. This better scaling is crucial for the applications targeted in this thesis, where we study systems with a relatively large number of degrees of freedom (up to 10^4 or 10^5) in the central device region. In addition, the key role played by resonant reflection in the majority of the applications means that the dwell-time for charges in the system is also large, which necessitates long simulation times. This combination of large system size and long simulation time means that direct Green's function techniques are unsuitable. It should be noted that for other applications, such as molecular electronics, the scaling properties of the method may not be so important and a direct Green's function technique may be a better choice.

References

1. S. Datta, Electronic transport in mesoscopic systems, in *Cambridge Studies in Semiconductor Physics and Microelectronic Engineering 3* (Cambridge University Press, Cambridge, 2009)
2. R.P. Feynman et al., Quantum mechanics. Nachdr, in *The Feynman Lectures on Physics 3* (Addison-Wesley, Reading, 2007)
3. M.A. Topinka et al., Coherent branched flow in a two-dimensional electron gas. Nature **410**(6825), 183–186 (2001)
4. J.P. Bergfield, M.A. Ratner, Forty years of molecular electronics: Non-equilibrium heat and charge transport at the nanoscale: forty years of molecular electronics. Physica Status Solidi (b) **250**(11), 2249–2266 (2013)
5. M.J. Iqbal et al., Robust recipe for low-resistance ohmic contacts to a two-dimensional electron gas in a GaAs/AlGaAs heterostructure (2014). arXiv:1407.4781
6. T. Bautze et al., Theoretical, numerical, and experimental study of a flying qubit electronic interferometer. Phys. Rev. B **89**(12), 125432 (2014)
7. D. Ferry, S.M. Goodnick, *Transport in Nanostructures* (Cambridge University Press, 1997)
8. P.K. Tien, J.P. Gordon, Multiphoton process observed in the interaction of microwave fields with the tunneling between superconductor films. Phys. Rev. **129**(2), 647–651 (1963)
9. B.D. Josephson, Possible new effects in superconductive tunnelling. Phys. Lett. **1**(7), 251–253 (1962)
10. K.K. Likharev, *Dynamics of Josephson Junctions and Circuits* (Gordon and Breach Science Publishers, New York, 1986)
11. H. Pothier et al., Single-electron pump based on charging effects. EPL **17**(3), 249 (1992)
12. P.W. Brouwer, Scattering approach to parametric pumping. Phys. Rev. B **58**(16), R10135–R10138 (1998)
13. G. Fève et al., An on-demand coherent single-electron source. Science **316**(5828), 1169–1172 (2007)
14. A. Mahé et al., Current correlations of an on-demand single-electron emitter. Phys. Rev. B **82**(20) (2010)
15. F.D. Parmentier et al., Current noise spectrum of a single-particle emitter: theory and experiment. Phys. Rev. B **85**(16) (2012)
16. E. Bocquillon et al., electron quantum optics: partitioning electrons one by one. Phys. Rev. Lett. **108**(19) (2012)
17. E. Bocquillon et al., Coherence and indistinguishability of single electrons emitted by independent sources. Science **339**(6123), 1054–1057 (2013)
18. J. Dubois et al., Integer and fractional charge Lorentzian voltage pulses analyzed in the framework of photon-assisted shot noise. Phys. Rev. B **88**(8) (2013)
19. J. Dubois et al., Minimal-excitation states for electron quantum optics using levitons. Nature **502**(7473), 659–663 (2013)
20. L.S. Levitov, H. Lee, G.B. Lesovik, Electron counting statistics and coherent states of electric current. J. Math. Phys. **37**(10), 4845–4866 (1996)
21. J. Keeling, I. Klich, L.S. Levitov, Minimal excitation states of electrons in one-dimensional wires. Phys. Rev. Lett. **97**(11) (2006)
22. T. Jullien et al., Quantum tomography of an electron. Nature **514**(7524), 603–607 (2014)
23. C. Grenier et al., Electron quantum optics in quantum hall edge channels. Mod. Phys. Lett. B **25**(12n13), 1053–1073 (2011)

24. D.P. DiVincenzo, Quantum computation. Science **270**(5234), 255–261 (1995)
25. B.M. Terhal, M.M. Wolf, A.C. Doherty, Quantum entanglement: a modern perspective. Phys. Today **56**(4), 46–52 (2003)
26. C.W.J. Beenakker et al., Proposal for production and detection of entangled electron-hole pairs in a degenerate electron gas. Phys. Rev. Lett. **91**(14) (2003)
27. J. Rammer, *Quantum Field Theory of Non-equilibrium States* (Cambridge University Press, Cambridge, 2007)
28. C. Caroli et al., Direct calculation of the tunneling current. J. Phys. C Solid State Phys. **4**(8), 916–929 (1971)
29. L.V. Keldysh, Diagram technique for nonequilibrium processes. JETP **20**(47), 1515 (1964)
30. H.M. Pastawski, Classical and quantum transport from generalized Landauer-Büttiker equations. II. Time-dependent resonant tunneling. Phys. Rev. B **46**(7), 4053–4070 (1992)
31. N.S. Wingreen, A.-P. Jauho, Y. Meir, Time-dependent transport through a mesoscopic structure. Phys. Rev. B **48**(11), 8487–8490 (1993)
32. A.-P. Jauho, N.S. Wingreen, Y. Meir, Time-dependent transport in interacting and noninteracting resonant-tunneling systems. Phys. Rev. B **50**(8), 5528–5544 (1994)
33. Y. Meir, N.S. Wingreen, Landauer formula for the current through an interacting electron region. Phys. Rev. Lett. **68**(16), 2512–2515 (1992)
34. J. Rammer, H. Smith, Quantum field-theoretical methods in transport theory of metals. Rev. Mod. Phys. **58**(2), 323–359 (1986)
35. Y. Zhu et al., Time-dependent quantum transport: direct analysis in the time domain. Phys. Rev. B **71**(7), 075317 (2005)
36. C. Eduardo, Cuansing and Gengchiau Liang. J. Appl. Phys. **110**(8), 083704 (2011)
37. A. Prociuk, B.D. Dunietz, Modeling time-dependent current through electronic open channels using a mixed time-frequency solution to the electronic equations of motion. Phys. Rev. B **78**(16), 165112 (2008)
38. A. Croy, U. Saalmann, Propagation scheme for nonequilibrium dynamics of electron transport in nanoscale devices. Phys. Rev. B **80**(24), 245311 (2009)
39. V. Moldoveanu, V. Gudmundsson, A. Manolescu, Transient regime in nonlinear transport through many-level quantum dots. Phys. Rev. B **76**(8), 085330 (2007)
40. D. Hou et al., Time-dependent transport: time domain recursively solving NEGF technique. Phys. E Low-Dimension. Syst. Nanostruct. **31**(2), 191–195 (2006)
41. L. Zhang, J. Chen, J. Wang, First-principles investigation of transient current in molecular devices by using complex absorbing potentials. Phys. Rev. B **87**(20), 205401 (2013)
42. Y. Wei, J. Wang, Current conserving nonequilibrium ac transport theory. Phys. Rev. B **79**(19), 195315 (2009)
43. J. Wang, Time-dependent quantum transport theory from non-equilibrium Green's function approach. J. Comput. Electron. **12**(3), 343–355 (2013)
44. D. Kienle, M. Vaidyanathan, F. Léonard, Self-consistent ac quantum transport using nonequilibrium Green functions. Phys. Rev. B **81**(11), 115455 (2010)
45. P. Myöhänen et al., A many-body approach to quantum transport dynamics: Initial correlations and memory effects. EPL (Europhys. Lett.) **84**(6), 67001 (2008)
46. C. Verdozzi, C.-O. Almbladh, Kadanoff-Baym dynamics of Hubbard clusters: performance of many-body schemes, correlation-induced damping and multiple steady and quasi-steady states. Phys. Rev. B **82**(15), 155108 (2010)
47. J. Maciejko, J. Wang, H. Guo, Time-dependent quantum transport far from equilibrium: an exact nonlinear response theory. Phys. Rev. B **74**(8), 085324 (2006)
48. R. Tuovinen et al., Time-dependent Landauer-Büttiker formula: application to transient dynamics in graphene nanoribbons. Phys. Rev. B **89**(8), 085131 (2014)
49. M. Ridley, A. MacKinnon, L. Kantorovich, Current through a multilead nanojunction in response to an arbitrary time-dependent bias. Phys. Rev. B **91**(12), 125433 (2015)
50. M. Ridley, A. MacKinnon, L. Kantorovich, Calculation of the current response in a nanojunction for an arbitrary time-dependent bias: application to the molecular wire. J. Phys. Conf. Ser. **696**(1), 012017 (2016)

51. K.T. Cheung et al., Order O(1) algorithm for first-principles transient current through open quantum systems (2016). arXiv:1602.01638
52. M. Cini, Time-dependent approach to electron transport through junctions: general theory and simple applications. Phys. Rev. B **22**(12), 5887–5899 (1980)
53. G. Stefanucci, C.-O. Almbladh, Time-dependent partition-free approach in resonant tunneling systems. Phys. Rev. B **69**(19), 195318 (2004)
54. E. Perfetto, G. Stefanucci, M. Cini, Time-dependent transport in grapheme nanoribbons. Phys. Rev. B **82**(3), 035446 (2010)
55. Benoit Gaury et al., Numerical simulations of time-resolved quantum electronics. Phys. Rep. **534**(1), 1–37 (2014)
56. G. Stefanucci et al., Time-dependent approach to electron pumping in open quantum systems. Phys. Rev. B **77**(7), 075339 (2008)
57. S. Kurth et al., Time-dependent quantum transport: a practical scheme using density functional theory. Phys. Rev. B **72**(3), 035308 (2005)
58. G. Stefanucci, E. Perfetto, M. Cini, Ultrafast manipulation of electron spins in a double quantum dot device: a real-time numerical and analytical study. Phys. Rev. B **78**(7), 075425 (2008)
59. X. Qian et al., Time-dependent density functional theory with ultrasoft pseudopotentials: real-time electron propagation across a molecular junction. Phys. Rev. B **73**(3), 035408 (2006)
60. Z. Zhou, S.-I. Chu, A time-dependent momentum-space density functional theoretical approach for electron transport dynamics in molecular devices. EPL (Europhys. Lett.) **88**(1), 17008 (2009)
61. P. Bokes, F. Corsetti, R.W. Godby, Stroboscopic wave-packet description of nonequilibrium many-electron problems. Phys. Rev. Lett. **101**(4), 046402 (2008)
62. M. Konôpka, P. Bokes, Wave-packet representation of leads for efficient simulations of time-dependent electronic transport. Phys. Rev. B **89**(12), 125424 (2014)
63. R. Baer et al., Ab initio study of the alternating current impedance of a molecular junction. J. Chem. Phys. **120**(7), 3387–3396 (2004)
64. B. Novakovic, G. Klimeck, Atomistic quantum transport approach to time-resolved device simulations, in *2015 International Conference on Simulation of Semiconductor Processes and Devices (SISPAD)*, September 2015, pp. 8–11

Chapter 3
Numerical Algorithms for Time-Resolved Quantum Transport

In Sect. 2.3.3 we saw that wavefunction-based methods are the current state of the art for the types of systems we want to study (a large number of degrees of freedom and long times). In this chapter we shall explain the algorithms used to compute time-dependent observables using our wavefunction-based "source-sink" algorithm.

3.1 A High-Level Overview of the Algorithm

We wish to consider a general class of problems consisting of a finite scattering region, S, coupled to periodic, semi-infinite leads L. Recalling from Sect. 2.3 that we only wish to describe systems without electron-electron correlations, we may write the Hamiltonian for such a system as

$$\hat{H}(t) = \underbrace{\sum_{i,j \in S} H_{ij}(t)\hat{c}_i^\dagger \hat{c}_j}_{\hat{H}^S(t)} + \underbrace{\sum_{i \in S, j \in L} H_{ij}(t)\hat{d}_i^\dagger \hat{c}_j}_{\hat{H}^T(t)} + \underbrace{\sum_{i,j \in L} H_{ij}\hat{d}_i^\dagger \hat{d}_j}_{\hat{H}^L}, \qquad (3.1)$$

where $\hat{c}_i^\dagger$ ($\hat{c}_j$) are the fermionic creation (annihilation) operators for a one-particle state on site i (j). A "site" may label position as well as other degrees of freedom such as spin or orbital angular momentum, although in specific cases we shall often make the distinction between spatial and internal degrees of freedom explicit. We will denote $\mathbf{H}(t)$ the (infinite) matrix with elements $H_{ij}(t)$. The Hamiltonian of the central region is fully general, however we restrict the leads to be time independent (in addition to being periodic and semi-infinite). Each lead *remains in its own thermal equilibrium at all times*, though the leads may be out-of-equilibrium with respect to one another (different chemical potentials and temperatures). In addition, we restrict the time-dependent perturbations to positive times, so that $\mathbf{H}(t \leq 0) = \mathbf{H}_0$. Note

© Springer International Publishing AG 2017
J. Weston, *Numerical Methods for Time-Resolved Quantum Nanoelectronics*,
Springer Theses, DOI 10.1007/978-3-319-63691-7_3

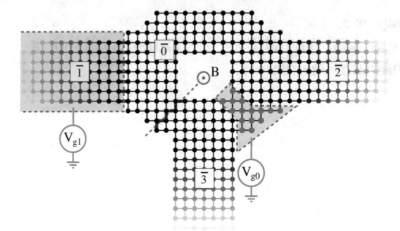

Fig. 3.1 Sketch of a typical system. It consists of a central scattering region, $\bar{0}$, attached to semi-infinite leads $\bar{1}$, $\bar{2}$, and $\bar{3}$. In this example there are two top gates with time-dependent voltages applied to them and a time-dependent magnetic field threaded through the central hole. The on-sites and hoppings affected by these time-dependent parameters are shown in **bold**. Note that in this gauge the gate applied to lead $\bar{1}$ affects the hoppings between the gate and the central region, $\bar{0}$ (see Appendix C)

that although this seems to put quite some restrictions on the class of systems we are considering, we can still handle:

- uniform, time-dependent voltages applied to leads (we need only perform a gauge transformation following the steps in Appendix C) and
- systems that start in a stationary, out-of-equilibrium state (i.e. different temperatures or chemical potentials in each lead).

Figure 3.1 shows an example of the type of system that we aim to simulate.

Before the time-dependent perturbations are switched on, the system is characterised by its scattering wavefunctions $\psi^{\text{st}}_{\alpha E}$, which are labelled by their energy E and incoming channel α. Explicitly, the $\psi^{\text{st}}_{\alpha E}$ are vectors of complex numbers that satisfy the eigenvalue equation

$$\mathbf{H}_0 \psi^{\text{st}}_{\alpha E} = E \psi^{\text{st}}_{\alpha E}. \tag{3.2}$$

The scattering wave functions $\psi^{\text{st}}_{\alpha E}$ are standard objects of mesoscopic physics and can be obtained directly by wave matching the incoming and outgoing modes at the lead-system boundary. An overview of the method for calculating the $\psi^{\text{st}}_{\alpha E}$ is given in Sect. 3.2. The expectation value of a physical observable $\hat{A} = \sum_{ij} A_{ij} \hat{c}^\dagger_i \hat{c}_j$, (e.g. electronic density or local currents) can be directly obtained by simply calculating the single-particle expectation values using these wavefunctions and weighting these according to Fermi statistics:

$$\langle \hat{A} \rangle = \sum_\alpha \int \frac{dE}{2\pi} \, f_\alpha(E) [\psi^{st}_{\alpha E}]^\dagger \mathbf{A} \, \psi^{st}_{\alpha E}, \tag{3.3}$$

where $f_\alpha(E)$ is the Fermi function of the electrode associated with channel α and $\mathbf{A}$ is the matrix with elements A_{ij}.

The generalisation of Eq. (3.3) to the time-dependent problem is rather straight-forward: one first obtains the scattering states and lets them evolve according to the time-dependent Schrödinger equation

$$i \frac{\partial}{\partial t} \psi_{\alpha E}(t) = \mathbf{H}(t) \psi_{\alpha E}(t) \tag{3.4}$$

with the initial condition $\psi_{\alpha E}(t = 0) = \psi^{st}_{\alpha E}$. The observables follow from Eq. (3.3) where the $\psi^{st}_{\alpha E}$ are replaced by $\psi_{\alpha E}(t)$:

$$\langle \hat{A}(t) \rangle = \sum_\alpha \int \frac{dE}{2\pi} \, f_\alpha(E) \psi^\dagger_{\alpha E}(t) \mathbf{A} \, \psi_{\alpha E}(t). \tag{3.5}$$

The equivalence between this prescription and the non-equilibrium Green's function formalism was derived in Ref. [1].

Common observables to calculate are the charge on a single site, $\hat{\rho}_i$, and the current flowing between two sites, $\hat{I}_{ij}$, which can be written

$$\langle \hat{\rho}_i(t) \rangle = q \sum_\alpha \int \frac{dE}{2\pi} \, f_\alpha(E) [\psi^\dagger_{\alpha E}(t)]_i \, [\psi_{\alpha E}(t)]_i \tag{3.6}$$

$$\langle \hat{I}_{ij}(t) \rangle = q \sum_\alpha \int \frac{dE}{2\pi} \, f_\alpha(E) 2\Im \left([\psi^\dagger_{\alpha E}(t)]_i \, H_{ij}(t) \, [\psi_{\alpha E}(t)]_j \right). \tag{3.7}$$

where $[\psi_{\alpha E}(t)]_i$ is the component of $\psi_{\alpha E}(t)$ on site i, $\Im$ is the imaginary part and q is the charge of the particles. The charge and current are related by a continuity equation:

$$\frac{\partial}{\partial t} \hat{\rho}_i(t) - \sum_j \hat{I}_{ij}(t) = 0, \tag{3.8}$$

which also defines the direction of the current.

3.1.1 A Note on Bound States

It should be noted that Eq. (3.5) does not account for the case where there are true bound states present in the system. These states are problematic as they do not hybridise with the continuum of states in the electrodes. The problem of including the contribution of bound states into non-equilibrium simulations has been discussed for some years now [2–6]. Without a unique (and known) distribution function for the whole system additional models are needed to describe how the bound states should be initially filled. The scattering states do not have this problem, as they originate deep in the electrodes, which are always in equilibrium (in our treatment). The role of bound states is particularly important for the treatment of Josephson junctions, where the presence of true Andreev bound states is responsible for the various Josephson effects (see Chap. 6). To circumvent this obstacle we use the procedure outlined in Ref. [7] when the system contains true bound states. The scheme consists of calculating the bound states for $t \leq 0$, ψ_{nE}^{bnd}, filling them up according to the Fermi-Dirac distribution and then evolving them using Eq. (3.4), calculating observables using a modified form of Eq. (3.5):

$$\langle \hat{A}(t) \rangle = \sum_{\alpha} \int \frac{dE}{2\pi} f(E) \psi_{\alpha E}^{\dagger}(t) \mathbf{A} \, \psi_{\alpha E}(t) + \sum_{n} f(E_n) [\psi_{nE}^{bnd}(t)]^{\dagger} \mathbf{A} \, \psi_{nE}^{bnd}(t).$$

(3.9)

Note that in order for this procedure to be valid, we must *start in equilibrium* at $t = 0$ so that there is a unique Fermi-Dirac distribution $f(E)$ for the system. If we do not start from equilibrium we must introduce additional models that will tell us how to correctly populate the bound states of the non-equilibrium system; while this physics is interesting, it is not the subject of this thesis. In any case, for $t > 0$ we can always apply a bias voltage to the electrodes to bring the system into a non-equilibrium situation. In this thesis presence of bound states is determined on a case-by-case basis, and in practice their energies and wavefunctions are determined by truncating the infinite system, so that it consists of the central region and a large number of layers in the electrodes, and diagonalising the resulting Hamiltonian. The band structure of the infinite electrodes is also calculated, and any eigenstates of the truncated system with energies outside the energy bands of the electrodes is classed as a bound state. The estimate for the bound states and their energies can be checked by truncating the system after an even larger number of electrode layers, and checking that the results match to within a certain tolerance.

3.1.2 Algorithm Overview

In essence, then, calculating time-resolved observables using our scheme consists of the following steps:

1. defining a tight-binding Hamiltonian;
2. calculating the scattering states at $t = 0$, before the time-dependent perturbations;
3. evolving the scattering states up to a time t;
4. calculating the single-particle expectation value of the observable for each scattering state;
5. performing the integral (3.5) to calculate the full thermal expectation value;
6. repeating from step 3 for all times for which we wish to calculate observables.

The definition of the tight-binding Hamiltonian and the calculation of the initial scattering states is handled by the KWANT package [8] (see Sect. 4.2 for details); we give an overview of the calculation of the stationary scattering states in Sect. 3.2. The algorithm for evolving the scattering states in time will be explained in Sect. 3.3, and the evaluation of the integral (3.5) will be discussed in Sect. 3.4.

3.2 Computing the Stationary Scattering States

In this section we will define the stationary scattering states in more precise terms and outline what equations one has to solve to obtain them. This is explained in more depth in Ref. [8], but is included here for completeness.

Firstly, we shall take the (infinite) Hamiltonian matrix for $t < 0$, $\mathbf{H}_0$, and express it in a more explicit form:

$$\mathbf{H}_0 = \begin{pmatrix} \ddots & \mathbf{V}_L & & \\ \mathbf{V}_L^\dagger & \mathbf{H}_L & \mathbf{V}_L & \\ & \mathbf{V}_L^\dagger & \mathbf{H}_L & \mathbf{V}_{LS} \\ & & \mathbf{V}_{LS}^\dagger & \mathbf{H}_S \end{pmatrix} \tag{3.10}$$

where $\mathbf{H}_S$ is the Hamiltonian of the central system (the part labelled $\bar{0}$ in Fig. 3.1), $\mathbf{H}_L$ is the Hamiltonian of a single unit cell of the semi-infinite leads (in the case where we have several physical leads we can always group them together into a single, virtual lead) and $\mathbf{V}_L$ is the hopping matrix between subsequent lead layers. $\mathbf{V}_{LS}$ is the hopping matrix between the central region and the first layer of the lead. We can also express $\psi_{\alpha E}^{st}$ in a more explicit form:

$$\psi_{\alpha E}^{st} = \begin{pmatrix} \vdots \\ \eta_{\alpha E}(2) \\ \eta_{\alpha E}(1) \\ \varphi_{\alpha E} \end{pmatrix} \tag{3.11}$$

where $\varphi_{\alpha E}$ is the scattering wavefunction in the central region, and $\eta_{\alpha E}(j)$ is the scattering wavefunction in slice j of the lead.

3.2.1 Calculation of the Lead Transverse Modes

To proceed we need to put some conditions on the form of the scattering wavefunction in the leads. To this end we will first need to treat the *infinite* lead in isolation (no coupling to the central region) in order to find the modes that it can support. We will see that the $\eta_{\alpha E}$ defined above will be a linear combination of these modes. Bloch's theorem applied to the infinite lead gives us a general form for the wavefunction in each cell, numbered by j:

$$\chi_{\alpha E}(j) = (\lambda_{\alpha E})^j \zeta_{\alpha E}. \tag{3.12}$$

The $\zeta_{\alpha E}$ satisfy

$$[\mathbf{H}_L + \lambda_{\alpha E}\mathbf{V}_L + \lambda_{\alpha E}^{-1}\mathbf{V}_L^\dagger]\zeta_{\alpha E} = E\zeta_{\alpha E}, \tag{3.13}$$

where we call $\zeta_{\alpha E}$ an incoming or outgoing *transverse mode* and $\lambda_{\alpha E}$ is an eigenvalue of the translation operator. $|\lambda_{\alpha E}| < 1$ corresponds to evanescent modes, while $\lambda_{\alpha E} = e^{ik_\alpha(E)}$ corresponds to propagating modes (with $k_\alpha(E)$ the longitudinal momentum of mode α). These propagating modes are normalised with respect to the expectation value of the current operator,

$$\langle I \rangle \equiv 2\Im[\chi_{\alpha E}^\dagger(j)\mathbf{V}_L\chi_{\alpha E}(j-1)] = \pm 1, \tag{3.14}$$

where $\Im$ is the imaginary part. This in turn allows us to identify three classes of modes: incoming $\chi_{\alpha E}^{in}$ ($\langle I \rangle = +1$), outgoing $\chi_{\alpha E}^{out}$ ($\langle I \rangle = -1$), and evanescent $\chi_{\alpha E}^{ev}$ ($\langle I \rangle = 0$). The incoming and outgoing modes have positive and negative group velocities, $v_{\alpha E}$, which can be calculated using the Hellman-Feynman theorem [9] and the relation $v_{\alpha E} = \partial E/\partial k_\alpha$:

$$v_{\alpha E} = i\zeta_{\alpha E}^\dagger(\mathbf{V}e^{ik_\alpha(E)} - \mathbf{V}^\dagger e^{-ik_\alpha(E)})\zeta_{\alpha E}. \tag{3.15}$$

To solve the (second order) Eq. (3.13), we can cast it into the form of a generalised eigenvalue problem:

$$\begin{pmatrix} \mathbf{H}_L - E\mathbf{1} & \mathbf{V}_L^\dagger \\ \mathbf{1} & \mathbf{0} \end{pmatrix} \begin{pmatrix} \zeta_{\alpha E} \\ \rho_{\alpha E} \end{pmatrix} = \lambda_\alpha \begin{pmatrix} -\mathbf{V}_L & \mathbf{0} \\ \mathbf{0} & \mathbf{1} \end{pmatrix} \begin{pmatrix} \zeta_{\alpha E} \\ \rho_{\alpha E} \end{pmatrix} \tag{3.16}$$

where $\rho_{\alpha E}$ is defined by the second line of Eq. (3.16). Efficient techniques have been developed to solve such problems [10, 11].

3.2.2 A Closed Set of Equations for the Scattering States

Using the above definitions we can now proceed to write a general form for the scattering wavefunction in the leads. In fact we *define* the scattering wavefunction in the lead to be a single incoming mode and a superposition of outgoing and evanescent modes:

$$\eta_{\alpha E}(j) = \chi_{\alpha E}^{in}(j) + \sum_\beta S_{\alpha\beta}\chi_{\beta E}^{out}(j) + \sum_\beta \tilde{S}_{\alpha\beta}\chi_{\beta E}^{ev}(j), \tag{3.17}$$

where $\chi_{\alpha E}^{in}$ ($\chi_{\alpha E}^{out}$) are the incoming (outgoing) modes (i.e. those with negative and positive velocity respectively), and the $S_{\alpha\beta}$ and $\tilde{S}_{\alpha\beta}$ are the scattering amplitudes from mode β to mode α. The $S_{\alpha\beta}$ are the elements of the so called *scattering matrix*. We thus see that the scattering states are *labelled by their incoming mode*, α, although they also in general have components in many outgoing (and evanescent) modes due to scattering in the central region.

By inserting this form of $\eta_{\alpha E}(j)$ into the infinite linear system, $\mathbf{H}_0 \psi_{\alpha E}^{st} = E \psi_{\alpha E}^{st}$, we are able to extract a closed set of equations for $\varphi_{\alpha E}$ (the scattering wavefunction inside the system) and the $S_{\alpha\beta}$ (the components of the scattering matrix). This closed set of linear equations is then solved numerically using efficient techniques [8, 12].

3.3 The Source-Sink Algorithm

In this section we will present the algorithm used for obtaining the time-evolved scattering states $\psi_{\alpha E}(t)$. We name this the "source-sink" algorithm due to the characteristic "source" and "sink" terms that appear in the time-dependent Schrödinger equation for the scattering states. We shall see that this algorithm is an improvement over the WF-C method of Ref. [1], as the latter scales as $\mathcal{O}(Nt_{max}^2)$, whereas the source-sink algorithm scales as $\mathcal{O}(Nt_{max})$ (N the size of the system and t_{max} the time up to which we wish to simulate). A large part of this section has been taken from Ref. [13].

3.3.1 The Source

In its original form, Eq. (3.4) is not very useful for numerics because the wavefunction spreads over the entire *infinite* system. A first simple, yet crucial, step consists of introducing the deviation $\bar{\psi}_{\alpha E}(t)$ from the stationary solution:

$$\psi_{\alpha E}(t) = e^{-iEt}[\psi_{\alpha E}^{st} + \bar{\psi}_{\alpha E}(t)]. \tag{3.18}$$

$\bar{\psi}_{\alpha E}(t)$ satisfies

$$i\frac{\partial}{\partial t}\bar{\psi}_{\alpha E}(t) = [\mathbf{H}(t) - E]\bar{\psi}_{\alpha E}(t) + S_{\alpha E}(t), \tag{3.19}$$

with

$$S_{\alpha E}(t) = [\mathbf{H}(t) - \mathbf{H}_0]\psi_{\alpha E}^{st} \tag{3.20}$$

and

$$\bar{\psi}_{\alpha E}(t = 0) = 0. \tag{3.21}$$

The new "source" term $S_{\alpha E}(t)$ can be computed from the knowledge of the stationary scattering states and is localised where the time-dependent perturbation takes place (where $H_{ij}(t) \neq H_{ij}(0)$). Equation (3.19) is already much better than Eq. (3.4) for numerics because the initial condition for $\bar{\psi}_{\alpha E}(t)$ vanishes everywhere. One can therefore truncate Eq. (3.19) and keep a finite system around the central time-dependent region where the source term lies. In practice, one adds N layers of each electrode. Note that in order for this procedure to be correct, the stationary scattering states are calculated for the *infinite* system and the truncation is only performed when evolving $\bar{\psi}_{\alpha E}(t)$ according to Eq. (3.19). For the truncation to be valid, the size of this added region must be larger than $N > v\, t_{max}/2$ where v is the maximum velocity at which the wave function can propagate and t_{max} the duration of the simulation. Hence, for large values of t_{max}, the total computational time to integrate Eq. (3.19) scales as $v\, t_{max}^2$. This algorithm corresponds to the WF-C algorithm of Ref. [1]. In Sect. 3.3.2 we will see that we can go beyond the WF-C method by introducing a complex absorbing potential in the region of added electrode. This will allow us to obtain $\mathcal{O}(N t_{max})$ scaling.

3.3.2 The Sink

The t_{max}^2 scaling of the WF-C algorithm comes from the fact that for long simulation times, one needs to introduce a large part of the leads ($\propto t_{max}$) in order to avoid spurious reflections at the boundaries where the leads have been truncated. To do any better than this, one needs to take advantage of the special structure of the leads: they are not only time-independent, but also invariant by translation. Hence, whatever enters into the lead will propagate towards infinity and never come back to the central region. Mathematically, the form of $\bar{\psi}_{\alpha E}(t)$ in the leads is a superposition of *outgoing* plane waves [1]:

$$\bar{\psi}_{\alpha E}(j, t) = \sum_{\alpha'} \int \frac{dE'}{2\pi} S_{\alpha'\alpha}(E', E) \chi_{\alpha'E'}^{out}(j) e^{-iE't} \qquad (3.22)$$

where $\bar{\psi}_{\alpha E}(j, t)$ is $\bar{\psi}_{\alpha E}(t)$ projected onto layer j of the lead, $\chi_{\alpha'E'}^{out}(j) \propto e^{ik_{\alpha'}(E')j}$ is defined in Sect. 3.2.1 and $S_{\alpha'\alpha}(E', E)$ is the time-dependent part of the inelastic scattering matrix. The crucial point of Eq. (3.22) is that it only contains outgoing modes, as the incoming one has been subtracted when removing the stationary scattering state. Therefore, once the wavefunction starts to reach the leads, it propagates toward infinity and never comes back to the central system.

A natural idea that comes to mind is to replace the finite fraction of the electrodes by some sort of (non-Hermitian) term in the Hamiltonian that "absorbs" the wavefunction that enters the leads. This has been studied in the literature in the context of various partial differential equations [14–20], as well as quantum transport more specifically [21, 22], and is usually known as a complex absorbing potential. The difficulty lies in the fact that this absorbing term must not give rise to reflections. At a given energy, a perfectly absorbing boundary condition does exist; it corresponds to adding the self energy of the lead at the boundary (which is a non-local complex absorbing potential, see WF-D method of Ref. [1]). However the outgoing waves of Eq. (3.22) span a finite energy window so that some energies would get reflected back to the central region. One solution to obtain a perfectly absorbing boundary condition is to use a boundary condition that is non local in time [14, 23]; this is the WF-B method of Ref. [1], and leads to algorithms that scale as t_{max}^2.

We choose instead to design an imaginary potential that varies spatially. We will show that *for any desired accuracy* we can design an imaginary potential that spreads over a finite width of N electrode unit cells—where N depends only on the required accuracy, not on t_{max}. In practice, this new algorithm is much more effective than WF-C when t_{max} becomes larger than the ballistic time of flight through the system. The idea behind the algorithm is fairly straightforward: suppose that a plane wave with a dispersion relation $E(k)$ propagates inside one electrode. If one adds an imaginary potential $-i\Sigma$ to the Schrödinger equation, this plane wave becomes evanescent which eventually leads to the absorption of the wave. On the other hand, any abrupt variation of this imaginary potential will lead to unwanted reflection back to the central part of the system. The imaginary potential must therefore be switched on *adiabatically* within a finite fraction of the electrodes, see Fig. 3.2 for a sketch. The new equation of motion contains both the previous source term and the additional sink in the electrodes,

$$i\frac{\partial}{\partial t}\bar{\psi}_{\alpha E}(t) = [\mathbf{H}(t) - E - i\Sigma]\bar{\psi}_{\alpha E}(t) + S_{\alpha E}(t), \qquad (3.23)$$

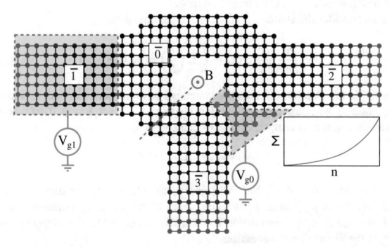

Fig. 3.2 Sketch of the truncated approximation to the system shown in Fig. 3.1, including the absorbing layers. The curve next to lead 2's absorbing layer shows a typical shape of the complex absorbing potential, $\Sigma(n)$

where the matrix Σ is diagonal and vanishes in the central region while it reads

$$\Sigma = \bigoplus_{n=1}^{n=N} \Sigma(n)\, \mathbf{1}_{\text{cell}} \tag{3.24}$$

in the absorbing layer placed at the beginning of the electrodes. The index n labels the unit cells of the leads, $\bigoplus$ is a direct sum, $\mathbf{1}_{\text{cell}}$ is the identity matrix defined over a unit cell, and $\Sigma(n)$ is a monotonically increasing scalar function. All that remains is to specify the function $\Sigma(n)$ so that it is of large enough magnitude to absorb all waves entering into the lead while being smooth enough so as not to produce spurious reflections. Our aim is to minimize the number of layers, N, that must be added in the system to absorb the outgoing waves without the error exceeding a tolerance δ.

3.3.3 Analytical Calculation of the Spurious Reflection

Before we can design a suitable imaginary potential, we must understand how the spurious reflection back to the central part depends on the shape of $\Sigma(n)$. We will start from a continuum model in order to develop an analytical solution for this simple case. The rationale, other than its tractability, is the fact that spurious reflections happen when $\Sigma(n)$ varies on a spatial scale that is short *compared to the wavelength* of the solution, hence it is dominated by small momentum k where the tight-binding dispersion relation reduces to its continuum limit. We will show that there is an extremely good agreement between the analytical results derived in this section and numerical calculations of the discretised model.

Let us consider the stationary 1D Schrödinger equation,

$$-\frac{\hbar^2}{2m^*}\frac{\partial^2}{\partial x^2}\psi(x) - \frac{i}{L}\Sigma\left(\frac{x}{L}\right)\psi(x) = E\psi(x) \tag{3.25}$$

where m^* is the electron effective mass and we have introduced a length scale, L, which controls how fast $\Sigma(x)$ varies. For negative x, we set $\Sigma(x \le 0) = 0$ so that the wavefunction is in a superposition of plane waves,

$$\psi(x) = e^{ikx} + r_\Sigma e^{-ikx} \tag{3.26}$$

where we *define* $E = \hbar^2 k^2 / 2m^*$. Our goal is to calculate the spurious reflection probability $R_\Sigma = |r_\Sigma|^2$ induced by the presence of the imaginary potential. We first rescale the equation by E and define $\bar{x} = kx$, $\bar{\Sigma}(u) = (k/E)\Sigma(u)$ and $\psi(x) = \bar{\psi}(\bar{x})$ to obtain the dimensionless equation,

$$\left[\frac{\partial^2}{\partial \bar{x}^2} + \frac{i}{kL}\bar{\Sigma}\left(\frac{\bar{x}}{kL}\right) + 1\right]\bar{\psi}(\bar{x}) = 0 \tag{3.27}$$

with

$$\bar{\psi}(\bar{x}) = e^{i\bar{x}} + r_{\Sigma} e^{-i\bar{x}} \tag{3.28}$$

for $\bar{x} < 0$. It is apparent from Eq. (3.27) that the spurious reflection is controlled by the dimensionless parameter kL. Since we want this spurious reflection to be small, we will work in the limit of large $kL \gg 1$ and expand r_{Σ} in powers of $1/kL$. The zeroth order contribution is simply the extension of the WKB limit to imaginary potential; the wavefunction takes the form of an evanescent wave,

$$\bar{\psi}(\bar{x}) \approx e^{\bar{S}(\bar{x})}, \tag{3.29}$$

with $\bar{S}(\bar{x})$ satisfying

$$[\bar{S}'(\bar{x})]^2 + 1 + i\frac{1}{kL}\bar{\Sigma}\left(\frac{\bar{x}}{kL}\right) = 0 \tag{3.30}$$

where primes denote derivatives. We expand $\bar{S}(\bar{x})$ to first order in $1/kL$, and apply the boundary condition Eq. (3.28) at $\bar{x} = 0$, as well as $\bar{\psi}(kL) = 0$ (perfect reflection at the end of the simulation domain at $x = L$) to obtain the zeroth order contribution to r_{Σ}:

$$r_{\Sigma}^0 = e^{2ikL}e^{-Ak/E}, \tag{3.31}$$

where

$$A = \int_0^L \frac{1}{L}\Sigma\left(\frac{x}{L}\right)\,\mathrm{d}x \tag{3.32}$$

is independent of kL. Physically speaking, the wavefunction is exponentially attenuated up to the hard wall at $x = L$ where it is fully reflected and then again exponentially attenuated until $x = 0$.

The contribution r_{Σ}^0 takes into account the finite absorption due to the imaginary potential, but not the spurious reflections due to wavevector mismatch. It is therefore necessary to go beyond the *adiabatic* WKB approximation and calculate its $1/kL$ deviation, r_{Σ}^1. We can ignore the hard wall at $x = L$ as it will play no role in what follows. Generalizing the WKB approximation we choose the following ansatz for $\bar{x} > 0$:

$$\bar{\psi}(\bar{x}) = \bar{\varphi}(\bar{x})e^{\bar{S}(\bar{x})} \tag{3.33}$$

$\bar{S}(\bar{x})$ contains the fast oscillating and decaying parts, while $\bar{\varphi}(\bar{x})$ contains the remaining (slow) parts. Plugging the ansatz Eq. (3.33) into Eq. (3.27), our Schrödinger equation becomes

$$\left\{ \bar{\varphi}''(\bar{x}) + \left[2i - \frac{1}{kL}\bar{\Sigma}\left(\frac{\bar{x}}{kL}\right) + 2\mathcal{O}\left(\frac{1}{(kL)^2}\right) \right]\bar{\varphi}'(\bar{x}) \right.$$
$$\left. + \left[\frac{-1}{2(kL)^2}\bar{\Sigma}'\left(\frac{\bar{x}}{kL}\right) + \mathcal{O}\left(\frac{1}{(kL)^3}\right) \right]\bar{\varphi}(\bar{x}) \right\}e^{\bar{S}(\bar{x})} = 0 \tag{3.34}$$

with

$$\bar{S}(\bar{x}) = i\bar{x} - \frac{1}{2} \int_0^{\bar{x}/kL} \bar{\Sigma}(u)\, du + \mathcal{O}\left(\frac{1}{kL}\right). \tag{3.35}$$

We write $\bar{\varphi}(\bar{x})$ as $\bar{\varphi}(\bar{x}) = \bar{\varphi}_0(\bar{x}) + (1/kL)\bar{\varphi}_1(\bar{x})$ and notice that, in the limit $(1/kL) \to 0$, Eq. (3.34) admits a solution $\bar{\varphi}(\bar{x}) = \bar{\varphi}_0(\bar{x}) = A + Be^{-2i\bar{x}}$. In this limit there should be no backscattering from the imaginary potential, so $B = 0$ and $\bar{\varphi}_0(\bar{x}) = 1$, to match the boundary conditions (3.28). The derivatives of $\bar{\varphi}_0(\bar{x})$ hence vanish and we arrive at

$$\bar{\varphi}_1''(\bar{x}) + 2\left[i - \frac{1}{2kL}\bar{\Sigma}(\bar{x}/kL)\right]\bar{\varphi}_1'(\bar{x}) = \frac{1}{2kL}\bar{\Sigma}'(\bar{x}/kL) \tag{3.36}$$

up to terms of order $\mathcal{O}(1/kL)^2$. Equation (3.36) can be solved by the variation of constant method,

$$\bar{\varphi}_1'(\bar{x}) = \bar{C}(\bar{x}) \exp\left\{-2i\bar{x} + \int_0^{\bar{x}/kL} \bar{\Sigma}(u)\, du\right\} \tag{3.37}$$

with

$$\bar{C}'(\bar{x}) = \frac{1}{2kL}\bar{\Sigma}'(\bar{x}/kL) \exp\left\{2i\bar{x} - \int_0^{\bar{x}/kL} \bar{\Sigma}(u)\, du\right\}. \tag{3.38}$$

Applying the continuity condition on $\bar{\psi}(\bar{x})$ and $\bar{\psi}'(\bar{x})$ at $\bar{x} = 0$ we obtain the 1st order contribution to the reflection amplitude:

$$r_\Sigma^1 = \frac{-1}{2ikL}\bar{C}(0), \tag{3.39}$$

which we can write explicitly, using Eq. (3.38) and the condition $\bar{C}(\infty) = 0$, as

$$r_\Sigma^1 = \frac{1}{4ikL} \int_0^\infty \bar{\Sigma}'(u) \exp\left\{2ikLu - \int_0^u \bar{\Sigma}(v)\, dv\right\} du. \tag{3.40}$$

One can understand r_Σ^1 as the Fourier transform at (large) frequency (kL) of the gradient of the imaginary potential weighted by the absorption that has already taken place. Putting together Eqs. (3.31) and (3.40), we finally obtain

$$\begin{aligned}
r_\Sigma = {}& e^{2ikL} e^{-Ak/E} \\
& + \frac{1}{4iEL} \int_0^\infty \Sigma'(u) \exp\left\{2ikLu - \frac{k}{E}\int_0^u \Sigma(v)dv\right\} du.
\end{aligned} \tag{3.41}$$

Equation (3.41) is the main result of this section. Now that we understand how the spurious reflection depends on the shape of $\Sigma(x)$, we need to design the imaginary potential so as to minimize Eq. (3.41) (for a given L). More precisely, for a given required precision ε, we wish to enforce $R_{\Sigma} < \varepsilon$ *irrespective* of the value of the energy E. Such a stringent condition is not, strictly speaking, feasible as $R_{\Sigma} \to 1$ when $E \to 0$ (all the variations of the imaginary potential become "abrupt" when the electronic wavelength becomes infinite), but we will see that the associated error for the time-dependent problem *can* be kept under control.

Now let us concentrate on an algebraic form for the imaginary potential:

$$\Sigma(u) = (n+1)Au^n, \tag{3.42}$$

from which the reflection amplitude calculated from Eq. (3.41) reads,

$$r_{\Sigma} = e^{2ikL}e^{-Ak/E} + \frac{An(n+1)(n-1)}{2^{n+2}Ek^nL^{n+1}} \tag{3.43}$$

As a consistency check of the approach developed above, we compare this analytical result for the reflection probability with direct numerical calculation using the KWANT d.c. transport package [8]. To do so we discretise the continuous Schrödinger equation onto a lattice of lattice spacing 1. Figure 3.3 shows how R_{Σ} scales for the case $n = 2$ and $n = 6$, showing an excellent agreement between the direct numerical simulations and the above analytical result in the limit of validity of the latter (small reflection). Figure 3.3c shows that the reflection has a minimum as a function of A which corresponds to a compromise between the first and last term of Eq. (3.43). Once A has been chosen large enough for the first term of Eq. (3.43) to be negligible, one can always choose L large enough to control the second term. We can already anticipate that the difficulties will come from vanishing energies $E \to 0$ for which the spurious reflection tends to unity.

3.3.4 Numerical Precision in the Time Domain

Now that we understand the d.c. case, let us consider the previous one dimensional model in the time domain and send a Gaussian voltage pulse through the wire. This problem has been studied in detailed in Ref. [1], to which we refer for more details on the physics. We compute the current flowing and measure the error with respect to a reference calculation $I_E^{\text{ref}}(t)$,

$$\delta = \frac{\int_0^{t_{\max}} |I_E(t) - I_E^{\text{ref}}(t)|dt}{\int_0^T |I_E^{\text{ref}}(t)|dt} \tag{3.44}$$

where $I_E(t)$ is the time-dependent probability current for a particle injected at energy E using the above designed imaginary potential to absorb the outgoing waves. The reference calculation is performed without imaginary potential, but with enough added lead cells such that the solution does not have time to propagate back into the

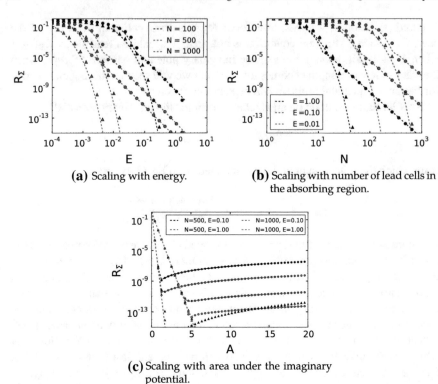

(a) Scaling with energy.

(b) Scaling with number of lead cells in the absorbing region.

(c) Scaling with area under the imaginary potential.

Fig. 3.3 d.c. reflection probability of a one dimensional chain in presence of an imaginary potential. The three panels show the scaling with different parameters. Symbols are numerical simulation of the discrete model and *dashed lines* are the analytic (continuum) result, Eq. (3.43), with $n = 2$ (*circles*) and $n = 6$ (*triangles*). N is the number of cells in the absorbing region

central region before the end of the simulation (i.e. the reference calculation uses the WF-C method of Ref. [1]).

Figure 3.4a shows the scaling of the error δ in the time-dependent calculation with respect to the d.c. reflection probability of the absorbing region R_Σ as N is changed. The current at an energy at the centre of the spectrum is calculated. We see from Fig. 3.4 that for very short absorbing regions the error scales proportionally to R_Σ, whereas for longer regions it scales as $\sqrt{R_\Sigma}$. This simply reflects the fact that the error on $\bar{\psi}_{\alpha E}(t)$ is proportional to $\sqrt{R_\Sigma} = r_\Sigma$; since the current (hence δ) is quadratic in $\psi_{\alpha E}(t) \equiv e^{-iEt}(\psi^{st}_{\alpha E} + \bar{\psi}_{\alpha E}(t))$, the error has the form $\delta \sim 2|\psi^{st}_{\alpha E}|\sqrt{R_\Sigma} + R_\Sigma$. More importantly, we see that we can control the error with arbitrary precision and for extremely long times (we checked this last point for much longer times than those shown in the inset).

More interesting is the behaviour of the error δ as a function of the injection energy E. Indeed, since $R_\Sigma \to 1$ when $E \to 0$, we might expect δ to behave badly as one decreases the energy. Figure 3.4b indeed shows that the error increases as the energy

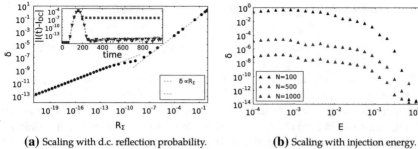

(a) Scaling with d.c. reflection probability.   (b) Scaling with injection energy.

Fig. 3.4 Scaling of the error, δ, in the time-dependent simulation with respect to: **a** the d.c. reflection probability, R_Σ, and **b** the particle injection energy, E. A monomial CAP with $n = 6$ was used. For (**a**) simulations were carried out at a single energy at the centre of the band and the length of the absorbing region was varied. *Dashed lines* show fits to $\delta \propto R_\Sigma$ (*blue*) and $\delta \propto \sqrt{R_\Sigma}$ (*red*). *Inset* Deviation of the probability current from equilibrium for different lengths of the absorbing region corresponding to the two points indicated by *arrows* in the main figure. The *black dashed curve* shows the reference calculation

is lowered, however, one finds that δ *saturates* at small energy. Furthermore, the saturated valued decreases with N and can thus be controlled. This behaviour comes from the structure of the wavefunction as shown in Eq. (3.22); even though one injects an electron at a definite energy inside the system, the energy of the outgoing wave is ill defined. The contribution to the wavefunction coming from spurious reflections takes the form

$$\delta\bar\psi_{\alpha E}(j, t) = \sum_{\alpha'} \int e^{-iE't}\chi^{\text{in}}_{\alpha' E'}(j)r_\Sigma(E')\, S_{\alpha'\alpha}(E', E)\mathrm{d}E' \qquad (3.45)$$

The contribution spreads over an energy window E_{pulse} which characterizes the inelastic scattering matrix, $S_{\alpha'\alpha}(E', E)$. $S_{\alpha'\alpha}(E', E)$ typically decays on an energy scale of the order of $E_{pulse} = \hbar/\tau_{pulse}$ (see Fig. 10 of Ref. [1] for an example). For the voltage pulse considered here (which sends one electron through the system), τ_{pulse} is essentially the duration of the pulse. The consequence is that the reflection r_Σ is averaged over an energy window of width E_{pulse}, which blurs the $E = 0$ behaviour of r_Σ:

$$\delta \approx \langle r_\Sigma(E)\rangle_{E<E_{pulse}} \approx r_\Sigma(E_{pulse}) \qquad (3.46)$$

We conclude that the error can always be made arbitrarily small, irrespective of the duration of the simulation. A slight drawback is that for a given imaginary potential, the precision of the calculation can depend on the actual physics taking place inside the central system (which sets E_{pulse}) if one injects electrons with energies close to the band edges of the leads.

3.3.5 Additional Ingredients for a Robust Algorithm

We now discuss how to turn the above results into a practical scheme to perform numerical calculations in a robust way.

Since we cannot guarantee the error for a given shape of the imaginary potential (we have seen above that it might depend on the physics of the central region), we first need to design an algorithm for an on-fly calculation of an error estimate without the reference calculation used above. This can be done as follows for a small additional computational cost. In the integration of the Schrödinger equation, one separates the wavefunction in the central region $\bar{\psi}_{\bar{0}}$ and in the leads $\bar{\psi}_{\bar{1}}$ (let us suppose that there is only one lead for simplicity). The equations to be integrated take the block form,

$$i\frac{\partial}{\partial t}\bar{\psi}_{\bar{0}} = \mathbf{H}_{\bar{0}\bar{0}}(t)\bar{\psi}_{\bar{0}} + \mathbf{H}_{\bar{0}\bar{1}}\bar{\psi}_{\bar{1}} + S_{\bar{0}}(t) \tag{3.47}$$

$$i\frac{\partial}{\partial t}\bar{\psi}_{\bar{1}} = \mathbf{H}_{\bar{1}\bar{1}}(\Sigma)\bar{\psi}_{\bar{1}} + \mathbf{H}_{\bar{1}\bar{0}}\bar{\psi}_{\bar{0}} \tag{3.48}$$

where $S_{\bar{0}}(t)$ is the source term present in the central region and the imaginary potential is included in $\mathbf{H}_{\bar{1}\bar{1}}$. One then introduces a second "copy" of the lead wavefunction $\bar{\psi}'_{\bar{1}}$ that uses a different imaginary potential $\mathbf{H}_{\bar{1}\bar{1}}(\Sigma')$. The equation of motion for this "copy" is

$$i\frac{\partial}{\partial t}\bar{\psi}'_{\bar{1}} = \mathbf{H}_{\bar{1}\bar{1}}(\Sigma')\bar{\psi}'_{\bar{1}} + \mathbf{H}_{\bar{1}\bar{0}}\bar{\psi}_{\bar{0}} \tag{3.49}$$

One then keeps track of both $\bar{\psi}_{\bar{1}}$ and $\bar{\psi}'_{\bar{1}}$ simultaneously, although only $\bar{\psi}_{\bar{1}}$ will affect the dynamics of $\bar{\psi}_{\bar{0}}$. The trick is to design $\Sigma'(n) = \Sigma(n - M)$, i.e. to insert M extra lead layers before the imaginary potential, and to monitor the difference between $\bar{\psi}'_{\bar{1}}$ and $\bar{\psi}_{\bar{1}}$ in the lead cell adjacent to the central region, $\delta\bar{\psi}_{\bar{1}} = \bar{\psi}_{\bar{1}} - \bar{\psi}'_{\bar{1}}$. Spurious reflections from the presence of Σ will arrive at the boundary of the central region for $\bar{\psi}_{\bar{1}}$ *before* $\bar{\psi}'_{\bar{1}}$, as the latter has M extra lead layers. This delay in the arrival of the spurious reflections will give rise to a finite $\delta\bar{\psi}_{\bar{1}}$. Note that $\delta\bar{\psi}_{\bar{1}}$ will remain 0 in the case that there are no spurious reflections. $\delta\bar{\psi}_{\bar{1}}$ can thus be used as an error estimate for the wave function in the lead.

In the worst case scenario this scheme will increase the computational cost by a factor of 2 (when the absorbing region represents the largest part of the system). It is worth noting, however, that without an error estimate for the spurious reflections one would have to check for convergence of results by performing several simulations with different values of N, the absorbing region length.

The remaining task is to choose the parameters A and N for a given shape of the imaginary potential. Ideally we would choose N as small as possible so as to minimize the extra computational effort while requiring that $|\delta\bar{\psi}_{\bar{1}}|$ remain smaller than a fixed maximum error, $\delta_{\max}$. Given $\delta_{\max}$ it is easy to choose A such that the first term in Eq. (3.43) is not a limitation. By noting that $e^{-Ak/E} < e^{-A/(a\gamma_B)}$ (where γ_B is the lead bandwidth and a is the discretisation step) we see that it is sufficient to choose

A such that $e^{-A/(aB)} < \delta_{\max}$ for the absorption process not to be the limiting factor of the precision. Next, one needs to choose N large enough to enforce $|\delta\bar{\psi_{\bar{1}}}| < \delta_{\max}$. This is currently tuned on a case-by-case basis; in practice, we found that a few hundred (up to a thousand) lead cells is almost always sufficient for the physics we have studied so far, for typical $\delta_{\max} \sim 10^{-5}$.

Let us end this Sect. 3.3.5 with a last point of practical importance. We have seen that the major contribution to spurious reflection comes from a narrow region around the band edge of the lead. The wavefunctions associated with these energies propagate *extremely* slowly into the absorbing region due to the vanishing velocity at the band edge. Unless one is interested in extremely long simulation times, we can take advantage of this by placing a small number of lead layers before the imaginary potential. The slow-moving waves will induce spurious reflections, but will take a long time to traverse this buffer layer due to their small group velocity. Meanwhile, the absorbing region does not have to be made as long, as it does not have to absorb waves of vanishingly small energy.

3.3.6 A Note on Timestepping Algorithms

All of the above treatment was essentially finding a coupled system of ODEs equivalent to Eq. (3.4) within a certain tolerance. Once we have constructed this "equivalent" system according to the above algorithm we still need to select a method to actually evolve the system forward in time. We choose to use the Runge-Kutta-Fehlberg (RKF45) method [24, 25], which is the stepper used in all the simulations presented in this thesis. Note, however, that the choice of stepper is an orthogonal concern to the source-sink algorithm, and our software implementation allows for any stepper to be used (see Sect. 4.2 for details), however the claimed $\mathcal{O}(Nt_{\max})$ complexity is, naturally, tied to the complexity of the stepper.

The RKF45 method is an embedded Runge-Kutta scheme that uses the same evaluation points to obtain an $\mathcal{O}(h^4)$ and an $\mathcal{O}(h^5)$ (h the timestep) accurate method. These estimates of different order can then be used to estimate the local error of the scheme, which can subsequently be used to control the size of the timestep. Such a scheme with an adaptive timestep is advantageous because it frees us from the need to pick a timestep that will be relevant at all energy scales (remember, we have to evolve $\psi_{\alpha E}(t)$ for a range of energies from the band edge up to the Fermi level). In addition, the use of an adaptive timestep means that even though the method is only conditionally stable (it is an explicit method), the algorithm will place the timestep within the radius of convergence.

3.3.6.1 The Unreasonable Efficacy of Runge-Kutta Methods

Runge-Kutta methods are not often used for the evolution of the time-dependent Schrödinger equation (TDSE) for two reasons: it is known to be stiff (and hence implicit methods may be more useful), and the norm of the solution must be preserved. Runge-Kutta methods are explicit (and so have a finite radius of convergence) and do not formally preserve the unitarity of the Schrödinger equation. Indeed, barring a few notable counterexamples [26–28] the most common method used seems to be Crank-Nicolson [23, 29, 30], however other approaches based on splitting the evolution operator [31–33] or approximating the evolution operator using a unitarity-preserving Magnus expansion appear to be popular [34–36]. The drawback of these more sophisticated methods are an increase in runtime and possibly algorithmic complexity. The use of implicit methods, for example, involves the solution of a (sparse) linear system of equations that scales polynomially with N—even with cutting-edge techniques [12]—compared to the $\mathcal{O}(N)$ complexity of the sparse matrix-vector products required for the explicit Runge-Kutta methods. The other methods mentioned invariably involve approximating some time ordered matrix exponential, which boils down to evaluating integrals of some commutator of sparse matrices. In addition, such methods are not amenable for calculation of an estimate of the local error, meaning that the timestep cannot be adapted as it can with the RKF45 method. On the other hand, we have found explicit methods to be more than adequate for all the problems that we have studied, despite the rich variety of physics involved: quantum Hall effect (see Sect. 4.3.2), superconductivity (see Chaps. 6 and 7), and floquet topological insulators (see Sect. 4.3.4) to name but a few.

Another argument that is commonly made against Runge-Kutta type steppers is that they do not respect the unitarity of the evolution one would expect from the Schrödinger equation. In practice for the transport problems studied in this thesis, however, this is not a concern for several reasons. Firstly, the presence of the sink term means that the evolution in the added lead cells is not unitary. Secondly, the equation of motion for $\bar{\psi}_{\alpha E}(t)$ also contains a *source* term, meaning that there is some $\bar{\psi}_{\alpha E}(t)$ "injected" on sites where there are time dependent perturbations. Both the above points stand irrespective of the stepping algorithm that is used. The only case where the lack of inherent unitarity in the stepping algorithm could be problematic is the case where the time-dependent perturbations happen in a finite time window and excite true bound states in the system. In this case, as the bound states do not hybridise with the leads, they can never relax and the wavefunction norm should be conserved in this case. If we were to perform simulations for extremely long times for such a case, the fact that the stepper does not preserve unitarity could become a problem.

3.4 Integrating the Observables

In this section we shall discuss method used for evaluating the integral (3.5). The first task is to perform a change of variables so that the integration variable is the quasi-momentum, k. This change of variables is advantageous because it is much easier to integrate numerically near the band edges in k-space. The reason for this is that the normalisation of the $\psi_{\alpha E}^{st}$ is such that they carry unit current (see Sect. 3.2.1). This means that they have an effective normalisation factor of $[v_\alpha(E)]^{-1/2}$, with $v_\alpha(k)$ the group velocity [37]. This in turn means that the quantity $[\psi_{\alpha E}^{st}]^\dagger \mathbf{A} \, \psi_{\alpha E}^{st}$ has an effective normalisation of $1/v_\alpha(E)$, which means that it will diverge whenever $v_\alpha(E) \to 0$. Even though this divergence is formally integrable ($v_\alpha(E) \propto \sqrt{E}$ for E close to a band edge), it renders a numerical evaluation more tricky. The dE/dk Jacobian factor (which is proportional to $v_\alpha(E)$) cancels with the $1/v_\alpha(E)$ when integrating over k, which regularises the integrand. We need only integrate in regions of k-space where there are modes *incoming* into the central region, which means $v_\alpha(k) > 0$ (as explained in Sects. 3.2.1 and 3.2.2). This gives us the following expression:

$$\langle \hat{A}(t) \rangle = \sum_\alpha \int_{-\pi}^{\pi} \frac{dk}{2\pi} \frac{dE}{dk} \, \Theta[v_\alpha(k)] \, f_\alpha(E(k)) \, \psi_{\alpha k}^\dagger(t) \mathbf{A} \, \psi_{\alpha k}(t), \qquad (3.50)$$

where $\Theta(x)$ is the Heaviside function.

Having obtained Eq. (3.50), we can proceed to use a quadrature scheme to numerically evaluate the individual integrals for each mode α. We typically use Gauss-Kronrod [38] nested quadrature schemes for this purpose. The advantage of nested quadrature schemes is that they allow an error estimate to be calculated by using a subset of the abscissae; no extra evaluations of the integrand are needed. This is particularly important for our application because an evaluation of the integrand at a given point k requires evolving $\psi_{\alpha k}$ in time numerically, which is a relatively expensive operation. In addition, Gauss-Kronrod rules (as opposed to other nested quadrature schemes such as Gauss-Lobatto) do not require the integrand to be evaluated on the boundary of the integration region. This is important, as the linear system to solve for $\psi_{\alpha k}^{st}$ is badly conditioned at points where $v_\alpha(k) = 0$ [37], and hence we cannot compute $\psi_{\alpha k}(t)$ at these points.

Initially we split the Brillouin zone into regions where $v_\alpha(k) < 0$ and $v_\alpha(k) > 0$ respectively. It is necessary to do this splitting for each mode independently, as in principle the modes can have very different dispersion relations. Because we are only interested in states corresponding to *incoming* modes, we need only integrate over regions where $v_\alpha(k) > 0$. This scheme is illustrated in Fig. 3.5a. In the case where there are no band crossings, the dispersion relation of each mode is C^1 continuous. It is therefore sufficient to find all local maxima/minima; this can be achieved using standard numerical routines [25, 39]. The case with band crossings is more complicated because the modes no longer have a consistent ordering in energy across the whole Brillouin zone, and so care is needed to properly identify the "same" mode when traversing a band crossing. This case has not been treated in detail, but would likely involve detecting where the band crossings occur and cutting the search intervals for minima of $E_\alpha(k)$ at these points. Band crossings could possibly be identified by searching for the roots of $\eta_{\alpha\beta}(k) = E_\alpha(k) - E_\beta(k)$.

Figure 3.5 compares the use of a k-space integration to a direct energy integration, as well as the use of a Gauss-Kronrod method rather than a simpler Simpson method [25], for a simple one-dimensional chain with a voltage pulse applied to the left lead. Figure 3.6 shows the same thing, but for a chain with some static disorder. The quantity calculated is the contribution to the current coming from the left lead

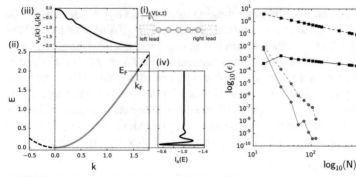

(a) (i) Sketch of the simulated setup: the solid line $V(t)$ shows the time-dependent voltage drop between the left lead and scattering region. (ii) The band structure in the left lead, with the integration region marked with a solid line. (iii) The integrand in momentum space, weighted by the mode velocity (see main text). (iv) The integrand in energy space.

(b) Scaling of the error ϵ in the numerical evaluation of the integral with the number of integrand evaluations N, for energy/momentum space integration using simpson/Gauss-Kronrod For the Gauss-Kronrod method the integration region was equally subdivided into $\lceil N/15 \rceil$ subregions and the 15-point rule was used on each of the subregions

Fig. 3.5 Comparison of different integration methods for the calculation of the current flowing through a one-dimensional (1D) chain after the application of a voltage pulse to the left lead

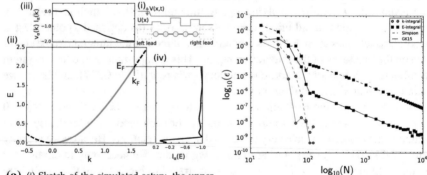

(a) (i) Sketch of the simulated setup: the upper solid line $V(t)$ shows the time-dependent voltage drop between the left lead and scattering region and the lower solid line $U(x)$ shows a typical static disorder potential in the scattering region. (ii) The band structure in the left lead, with the integration region marked with a solid line. (iii) The integrand in momentum space, weighted by the mode velocity (see main text). (iv) The integrand in energy space.

(b) Scaling of the error ϵ in the numerical evaluation of the integral with the number of integrand evaluations N, for energy/momentum space integration using simpson/Gauss-Kronrod For the Gauss-Kronrod method the integration region was equally subdivided into $\lceil N/15 \rceil$ subregions and the 15-point rule was used on each of the subregions

Fig. 3.6 Comparison of different integration methods for the calculation of the current flowing through a one-dimensional (1D) chain after the application of a voltage pulse to the left lead. The scattering region consists of 10 sites, and there static disorder in the onsite potentials in the range $[0, 0.06B]$ where B is the bandwidth

(i.e. a single term of the sum over α of Eqs. (3.5) and (3.50). In the system without disorder—where there is no backscattering—the divergence in the energy-space integrand can be clearly seen, and the error scaling of the corresponding integral (Fig. 3.5b, black squares and lines) is correspondingly unsatisfactory. In this case the momentum-space integral (red circles and lines) clearly has an advantage due to the regularised integrand. For the disordered system, however, the situation is a little different. The energy-space integrand no longer exhibits a strong divergence and the corresponding error scaling is much improved (although the momentum-space integral still has a clear advantage, especially for high-precision calculations). The reason for this is that the disorder introduces some finite backscattering into the system, which is especially strong for the modes near the band edge due to their low kinetic energy. The consequence of this is that the diverging contribution from these modes entering the system is compensated by their reflected components, giving a net contribution of (nearly) 0. We can thus see that while the momentum-space integration may not be "essential" for a large class of systems (those where the divergence is "naturally" regularised by backscattering processes), it nevertheless adds a degree of robustness to the integration, as it does not depend on the physics of the system being simulated.

After the initial integration intervals have been identified, the k-points corresponding to the abscissae of the 15(7)-point Gauss-Kronrod rule in each interval are computed. For each k-point we compute $\psi_{\alpha k}^{st}$ using the approach outlined in Sect. 3.2. We can now evolve all the $\psi_{\alpha k}(t)$ up to the first time τ_1 where we wish to calculate an observable. We then compute the integral (and error estimate) over each interval of k-points independently, using the Gauss-Kronrod weights. The total integral is then the sum of the integrals on each interval. The $\psi_{\alpha k}(t)$ can then be evolved to τ_2, the next time at which we wish to calculate an observable. This procedure is repeated until we have reached the maximum time t_{max} up to which we wish to simulate. Note that each interval of the integral can be evaluated independently, which lends itself to a trivial parallelisation of the algorithm. Additionally, we could use an *adaptive* integration, whereby If any interval has an error greater than a certain threshold then that interval is bisected, and the $\psi_{\alpha k}(t)$ for the k-points in these two new intervals are evolved from $t = 0$ to the time of interest. This procedure would continue until the computed value of the integral on each interval converges, and then the $\psi_{ak}(t)$ would be evolved to the next time of interest, where more subdivisions would possibly take place.

References

1. B. Gaury et al., Numerical simulations of time-resolved quantum electronics. Phys. Rep. **534**(1), 1–37 (2014)
2. R. Li et al., A corrected NEGF+DFT approach for calculating electronic transport through molecular devices: filling bound states and patching the non-equilibrium integration. Chem. Phys. **336**(2–3), 127–135 (2007)

3. A. Dhar, D. Sen, Nonequilibrium Green's function formalism and the problem of bound states. Phys. Rev. B **73**(8), 085119 (2006)
4. G. Stefanucci, Bound states in ab initio approaches to quantum transport: a time-dependent formulation. Phys. Rev. B **75**(19), 195115 (2007)
5. E. Khosravi et al., Bound states in time-dependent quantum transport: oscillations and memory effects in current and density. Phys. Chem. Chem. Phys. **11**(22), 4535–4538 (2009)
6. E. Khosravi et al., The role of bound states in time-dependent quantum transport. Appl. Phys. A **93**(2), 355–364 (2008)
7. R.E.V. Profumo et al., Quantum Monte Carlo for correlated out-of-equilibrium nanoelectronic devices. Phys. Rev. B **91**(24), 245154 (2015)
8. C.W. Groth et al., New J. Phys. **16**(6), 063065 (2014)
9. R.P. Feynman, Forces in molecules. Phys. Rev. **56**(4), 340–343 (1939)
10. M. Wimmer, Quantum transport in nanostructures: from computational concepts to spintronics in graphene and magnetic tunnel junctions. 1. Aufl. Dissertationsreihe der Fakultät für Physik der Universität Regensburg 5 (Univ.-Verl. Regensburg, Regensburg, 2009)
11. I. Rungger, S. Sanvito, Algorithm for the construction of self-energies for electronic transport calculations based on singularity elimination and singular value decomposition. Phys. Rev. B **78**(3), 035407 (2008)
12. P. Amestoy et al., A fully asynchronous multifrontal solver using distributed dynamic scheduling. SIAM. J. Matrix Anal. Appl. **23**(1), 15–41 (2001)
13. J. Weston, X. Waintal, A linear-scaling source-sink algorithm for simulating timeresolved quantum transport and superconductivity (2015). arXiv:1510.05967 [cond-mat]
14. X. Antoine et al., A review of transparent and artificial boundary conditions techniques for linear and nonlinear Schrödinger equations. Commun. Comput. Phys. **4**(4), 729–796 (2008)
15. J.G. Muga et al., Complex absorbing potentials. Phys. Rep. **395**(6), 357–426 (2004)
16. O. Shemer, D. Brisker, N. Moiseyev, Optimal reflection-free complex absorbing potentials for quantum propagation of wave packets. Phys. Rev. A **71**(3), 032716 (2005)
17. U.V. Riss, H.-D. Meyer, The transformative complex absorbing potential method: a bridge between complex absorbing potentials and smooth exterior scaling. J. Phys. B: At. Mol. Opt. Phys. **31**(10), 2279 (1998)
18. U.V. Riss, H.-D. Meyer, Reflection-free complex absorbing potentials. J. Phys. B: At. Mol. Opt. Phys. **28**(8), 1475 (1995)
19. D.J. Kalita, K. Ashish, J. Chem. Phys. **134**(9), 094301 (2011)
20. Z.H. Zhang, Use of negative complex potential as absorbing potential. J. Chem. Phys. **108**(4), 1429–1433 (1998)
21. R. Baer et al., Ab initio study of the alternating current impedance of a molecular junction. J. Chem. Phys. **120**(7), 3387–3396 (2004)
22. L. Zhang, J. Chen, J. Wang, First-principles investigation of transient current in molecular devices by using complex absorbing potentials. Phys. Rev. B **87**(20), 205401 (2013)
23. S. Kurth et al., Time-dependent quantum transport: a practical scheme using density functional theory. Phys. Rev. B **72**(3), 035308 (2005)
24. D.E. Fehlberg, Klassische Runge-Kutta-Formeln vierter und niedrigerer Ordnung mit Schrittweiten- Kontrolle und ihre Anwendung auf Wärmeleitungsprobleme. Computing **6**(1–2), 61–71 (1970)
25. W.H. Press (ed.), *Numerical Recipes: The Art of Scientific Computing*, 3rd edn (Cambridge University Press, Cambridge, 2007)
26. S.O. Fatunla, Numerical integrators for stiff and highly oscillatory differential equations. Math. Comput. **34**(150), 373–390 (1980)
27. J. Madroñero, B. Piraux, Explicit time-propagation method to treat the dynamics of driven complex systems. Phys. Rev. A **80**(3), 033409 (2009)
28. A.L. Frapiccini et al., Explicit schemes for time propagating many-body wave functions. Phys. Rev. A **89**(2), 023418 (2014)
29. J. Crank, P. Nicolson, A practical method for numerical evaluation of solutions of partial differential equations of the heat-conduction type. Math. Proc. Cambridge Philos. Soc. **43**(01), 50–67 (1947)

30. G. Stefanucci, E. Perfetto, M. Cini, Time-dependent quantum transport with superconducting leads: a discrete-basis Kohn-Sham formulation and propagation scheme. Phys. Rev. B **81**(11), 115446 (2010)
31. S. Blanes, P.C. Moan, Splitting methods for the time-dependent Schrödinger equation. Phys. Lett. A **265**(1–2), 35–42 (2000)
32. M. Thalhammer, High-order exponential operator splitting methods for time-dependent Schrödinger equations. SIAM J. Numer. Anal. **46**(4), 2022–2038 (2008)
33. B.A. Shadwick, W.F. Buell, Unitary integration with operator splitting for weakly dissipative systems. J. Phys. A: Math. Gen. **34**(22), 4771 (2001)
34. W. Magnus, On the exponential solution of differential equations for a linear operator. Comm. Pure Appl. Math. **7**(4), 649–673 (1954)
35. M. Hochbruck, C. Lubich, On Magnus integrators for time-dependent Schrödinger equations. SIAM J. Numer. Anal. **41**(3), 945–963 (2003)
36. S. Blanes et al., The Magnus expansion and some of its applications. Phys. Rep. **470**(5–6), 151–238 (2009)
37. C. Groth et al., *Algorithms for Quantum Transport* (To be published) (2016)
38. R. Piessens et al., *Quadpack*. Springer Series in Computational Mathematics, vol. 1 (Springer, Berlin, Heidelberg, 1983)
39. R.P. Brent, *Algorithms for Minimization Without Derivatives*. Prentice-Hall series in Automatic Computation (Prentice-Hall, Englewood Cliffs, 1972)

Chapter 4
Software Design

In Chap. 3 the main algorithms used to simulate time-resolved quantum transport were presented. While the algorithms themselves are certainly important, one could argue that of almost equal importance is a concrete software implementation that actually delivers demonstrable added value to research projects. In this chapter we shall start by discussing the requirements and "philosophy" for good scientific software in the context of exploratory research. We will then move on to identify the abstractions that allow one to easily express the necessary concepts for time-resolved transport. In addition we shall identify the state of our current implementation, TKWANT, and what needs to be improved before it is ready for public release. Finally we will end with a short gallery of examples of TKWANT usage from projects that were mainly the work of collaborators, or where I personally had a role geared more towards the software than the physics. This is to show that despite its warts, the current implementation of TKWANT has been used to study a wide variety of physics outside the main applications that will be presented in part II of this thesis, and has achieved the goal of delivering demonstrable added value to research projects.

4.1 Requirements for Well Designed Scientific Software

The latter half of the twentieth century saw massive advances in both numerical algorithms and the computing power required to execute them. Indeed, while numerical methods for solving mathematical problems are not new (methods developed by Newton, Euler, and Gauss—to name but a few—are at the core of many modern algorithms), their great utility was only really unlocked at the advent of the digital computer. What has also changed to some extent is the relationship between numerical methods and their use in fundamental and applied research, especially in physics.

© Springer International Publishing AG 2017
J. Weston, *Numerical Methods for Time-Resolved Quantum Nanoelectronics*,
Springer Theses, DOI 10.1007/978-3-319-63691-7_4

One could say that the evolution of the physicist's use of numerical methods parallels the evolution of the computer programmer's use of the hardware at her disposal. Similarly to how the first general purpose, programmable computers were programmed directly in machine language,[1] the first uses of numerical methods by physicists were to solve problems of very restricted scope: evaluating a specific integral, or solving a particular differential equation. The appearance of "higher level" programming languages freed the programmer from the tedious bookkeeping required to write in machine language, and allowed her to more easily step back and see the big picture. Similarly, as more general purpose implementations of algorithms were developed, the physicist was freed from worrying about the implementation details and could instead concentrate on what matters: the physics. This development has continued up until the present day, where very high-level computing languages such as Python [1] and Ruby [2] almost totally obviate the need to worry about the internals of the computer. Similarly, in physics, there is a proliferation of black-box simulation software—especially in the fields of computational chemistry and nanoscience [3]—that obviates much of the need to understand the fundamental algorithms used.[2] While the parallel advances in programming languages and physics software have brought many positives, there are also negative consequences. In particular, high-level programming languages tend to produce programs that run more slowly,[3] and black-box physics software hides perhaps too much from the researcher using it.

In my opinion there is a missing abstraction layer for quantum transport in the current offering of software tools. At the lowest layer of abstraction there are a very solid set of interfaces for solving specific mathematical problems, such as BLAS and LAPACK for dense linear algebra [6, 7], or QUADPACK for numerical integration [8]. At the other end of the spectrum there are the professionally developed (and often non-free[4]) tools [10–12] that provide an enormous amount of functionality, but are often too rigid for exploratory research. In the middle there are a whole host of "home made" tools that are often used by only a single research group and passed down from student to student, each generation making their contribution: adding support for spin, allowing for a ring geometry with a hole, etc. The consequence of this is that there is an enormous amount of replicated effort. There is a real need to create an abstraction layer for the fundamental objects of quantum transport: Hamiltonians, Green's functions, scattering matrices, etc., that empowers researchers to easily set up and solve well-defined physical problems without having to be concerned about

---

[1]Machine language is the raw sequences of numbers that are interpreted and executed by the computer's processor.

[2]Recent technological advances even obviate the need to write a scientific article [4] and get it published in a peer review journal [5], but I digress.

[3]A function call in Python is roughly 100 times slower than a function call in the C language.

[4]I use the word "free" here, as defined by the Free Software Foundation [9], to refer to software that respects users' freedoms, rather than software that is *gratis*.

the low-level details. The researcher nevertheless needs to have the freedom to be able to set up *exactly* the calculation that they want, so such an abstraction layer should not "hard code" assumptions about the physics being studied.

4.1.1 KWANT: *An Example of Well Designed Scientific Software*

The KWANT package aims to be this missing layer in the software stack. KWANT essentially provides an interface for specifying a tight-binding Hamiltonian and computing quantities relevant for d.c. transport such as band structures, scattering matrices, and Green's functions. Notably, KWANT leaves the definition of the tight-binding Hamiltonian to the user; whether such a description was arrived at by discretising a continuum model or extracting tight binding parameters from a density functional theory calculation, is largely immaterial from KWANT's point of view.

Another notable difference between KWANT and traditional scientific codes is that KWANT is actually a *library* for the Python programming language, which means that simulations that use KWANT are written as small Python programs. This approach has several distinct advantages over the more traditional approach to scientific software, where a monolithic binary executable takes an input file, performs a computation, and produces an output file. Firstly, the Python programming language is simple to learn and offers almost infinite extensibility. Compare this to an input file approach where as soon as you want to express anything reasonably complex you have to essentially invent your own domain specific language in order to do it; re-implementing elementary concepts like variables, loops, and sequences. Secondly, there is an enormous ecosystem of extensions to the Python language [13] (called "packages"), that allow one to efficiently manipulate numeric arrays [14], do symbolic algebra operations [15], or produce publication-quality graphics [16]. This means that KWANT users (as well as KWANT developers) can leverage all this power in their scripts and simulations. Finally, while Python is a somewhat slow language it is also straightforward to write extension modules in low-level and fast languages like C and Fortran [17], which means that KWANT does not have to sacrifice run-time speed in the name of ease of use.

4.1.1.1 A Simple Example of KWANT Usage

To give more of a feel for how KWANT is actually used we will now go through a simple example of a quantum wire with a variable-height insulating barrier. We shall identify the three major steps in a KWANT simulation: system definition; finalisation of the system into a low-level format for efficient computation; and solution of the scattering problem. This will allow us to more easily appreciate the changes needed to incorporate time-resolved transport. This example should be comprehensible to readers who do not know the Python language, as long as they are familiar with programming concepts such as loops and subroutines (functions). Reference [18] provides a more complex example for the interested reader.

The first thing we need to do is to make KWANT available and define our lattice and an empty tight-binding system (any text on a line after a # is a comment):

```
import numpy  # useful numerical utilities
from matplotlib import pyplot  # plotting library
import kwant  # make Kwant available

lat = kwant.lattice.square()  # a square lattice for our sites
syst = kwant.Builder()  # an empty tight-binding system
```

Next, we need to define our tight-binding model, which we obtain from discretising a continuum Hamiltonian onto a square lattice following appendix A. First we define the scattering region, which is 20 sites long and 10 sites wide and contains the potential barrier (3 sites wide) that will induce backscattering:

```
# populate the system with sites, and set the on-site
# Hamiltonian matrix elements to +4
for x in range(20):
    for y in range(10):
        syst[lat(x, y)] = 4

# set all the nearest-neighbour hoppings to -1
syst[lat.neighbors()] = -1

# define a function for the insulating barrier
def insulating_barrier(site, Vg):
    return 4 + Vg

# change onsite matrix elements for all sites under the barrier
for x in range(9, 12):
    for y in range(10):
        syst[lat(x, y)] = insulating_barrier
```

We see that we can set Hamiltonian matrix elements in two ways: with explicit values, and with *functions*. The functions will be evaluated at the last possible moment, when the Hamiltonian is requested; this allows for a system to be constructed once at the start of the script, and then solved many times for different values of the system parameters. In the above example there is a single parameter Vg, which controls the onsite matrix element for a strip three sites wide in the centre of the scattering region.

Next we define the electrodes as tight-binding systems with a translational symmetry in the x direction, and we attach them to the scattering region:

```
lead = kwant.Builder(kwant.TranslationalSymmetry((-1, 0)))
for y in range(10):
    lead[lat(0, y)] = 4
lead[lat.neighbors()] = -1

# attach semi-infinite leads to the scattering region
syst.attach_lead(lead)  # lead from the left
syst.attach_lead(lead.reversed())  # lead from the right
```

Next we convert the system into a low-level format suitable for numerics:

```
fsyst = syst.finalized()
```

The consequence of this is that the geometry (or in other words the sparsity structure of the Hamiltonian) is fixed.

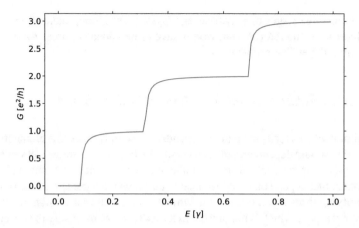

Fig. 4.1 Differential conductance through a quasi one-dimensional wire with an insulating barrier. This exact figure was produced from the script shown in the main text; an illustration of the simplicity of the KWANT package

Finally we are ready to compute the scattering matrix and transmission for the system:

```
Vg = 0.05
energies = numpy.arange(0, 1, 0.01)  # [0, 0.01, ..., 0.99]
transmissions = []
for E in energies:
    # calculate the scattering matrix at a given energy
    # with a given value of the system parameter(s) (args)
    smatrix = kwant.smatrix(fsyst, energy=E, args=(Vg,))
    T01 = smatrix.transmission(1, 0)  # lead 0 -> lead 1
    transmissions.append(T01)

pyplot.plot(energies, transmissions)  # plot the results
pyplot.xlabel(r'$E\_\[\gamma]$')  # set x-axis label
pyplot.ylabel(r'$G\_\[e^2/h]$')  # set y-axis label
pyplot.show()  # show the plot
```

This script produces the plot of differential conductance as a function of energy shown in Fig. 4.1. So we see that in 46 lines of simple code (the line count could be drastically reduced with more terse code) we can reproduce an elementary result of quantum transport: conductance quantisation.

4.2 TKWANT: Time-Dependent Extensions to KWANT

In its current iteration KWANT has no facilities for simulating time-dependent quantum transport. Part of the work of the past 3 years has been not only to develop the algorithms necessary for an efficient implementation of time-dependent quantum transport, but to identify the correct abstractions that will allow time-dependent transport to be added seamlessly to KWANT. In the following paragraphs we will discuss how the three stages of KWANT usage (system definition, finalisation, solving) would have to be altered (or what would have to be added) in order for KWANT to support time-dependent problems.

4.2.1 Modifications to the Problem Definition Stage

The addition when specifying a time-dependent system is that some elements of the Hamiltonian should depend on time. As we saw in the above example KWANT already supports assigning functions to Hamiltonian elements, so naturally any parts of the Hamiltonian that depend on time just need to take a `time` parameter. This is the way that specifying time-dependence is implemented in the current version of TKWANT.

A second aspect, which is not present in KWANT, is how to specify the observables that should be calculated during the solving phase. To be generally useful one should be able to supply a function that receives the KWANT system, the current time, and the scattering state wavefunction as inputs, and returns the action of the observable on the wavefunction. This is the interface used by the current version of TKWANT. Although such an interface is sufficiently general to capture all use cases, it is actually too general to be useful for most users. One particularly useful specialisation would be the case of general *densities* and their associated *currents*. We can define a density associated with a site i as

$$\rho_i^M(t) = \psi_i^\dagger(t) \mathbf{M}_i(t) \psi_i(t) \tag{4.1}$$

where $\psi_i(t)$ is a vector of wavefunction components on site i (e.g. spin or particle/hole degrees of freedom), and $\mathbf{M}_i(t)$ is a (possibly time-dependent) Hermitian matrix associated with site i. The associated current between sites i and j can be written

$$J_{ij}^M = 2\Im\{\psi_i^\dagger(t) \mathbf{M}_i(t) \mathbf{H}_{ij}(t) \psi_j(t)]\} \tag{4.2}$$

In the case of spin, for example, the $\mathbf{M}_i$ could be Pauli matrices, and the ρ_j^M would therefore be a "spin density". In order for a user to specify their density observable it would therefore be sufficient to provide a function that takes a site and a time and returns the appropriate Hermitian matrix representing the density for that site. This would be similar to the way in which the system Hamiltonian is defined with functions that take sites, as well as other parameters, and returns the Hamiltonian matrix elements. Building observables in this way could also be useful for KWANT users irrespective of TKWANT.

4.2.2 Modifications to the Finalisation Stage

The KWANT low-level system format allows for the Hamiltonian to be evaluated, and for the structure of the system to be efficiently queried (e.g. which sites are connected via hoppings etc.). TKWANT requires an additional component: evaluation of the time-dependent part of the Hamiltonian. Often the time dependence of a system will only affect a limited number of matrix elements, and in such cases it is wasteful to re-evaluate the full Hamiltonian. The low-level format required by TKWANT therefore needs an efficient way to query the system as to which matrix elements depend on time, and only re-evaluate these. In the current version of TKWANT a new type of low-level system is used for this purpose, and KWANT systems are finalised into this new "TKWANT finalised system".

4.2.3 Modifications to the Solving Stage

The solving stage will clearly be vastly different for TKWANT when compared with KWANT. While the KWANT solvers all boil down essentially to solving a linear system (scattering matrix and scattering wavefunction calculations) or an eigenvalue problem (calculating band structure), the source-sink algorithm at the heart of TKWANT requires evolving many differential equations (one for each mode and value of momentum/energy) and numerically integrating the results. The TKWANT solver will therefore be completely independent from the other KWANT solvers, but will need to use them in order to obtain the initial conditions for the scattering states. A user of the solver should simply be able to provide: a system that depends on time, a list of observables to calculate, and the times at which to evaluate them, and get back the thermal average of the observables evaluated at those times. One challenge to overcome is that there are many more parameters associated with the source-sink algorithm: what boundary conditions to use, what algorithm to use to evolve the scattering states, and what algorithm to use for the integrals over momentum/energy. There is also the question of how errors will accumulate during the calculation. The choices made for the current implementation were discussed in Sects. 3.3.5, 3.3.6 and 3.4. A future version of TKWANT should certainly allow for arbitrary combinations of boundary conditions/steppers/integrators.

Another challenge is that an efficient implementation of the source-sink algorithm will necessarily involve parallel computations, as each $\psi_{\alpha k}(t)$ evolves independently of the others. The current version of TKWANT exploits this available parallelism, however it does not implement an adaptive energy/momentum integration (see the last paragraph of Sect. 3.4). Instead, one merely chooses a set of integration regions at the start of the calculation and the observables (and an error estimate) are calculated using these regions during the whole calculation. This means that the convergence (or not) of the calculation can only be verified at the end. While this mode of operation is useful when a given set of integration regions is known to produce results that converge, it is particularly inefficient when no knowledge of the structure of the observables in energy/momentum is known. In this (common) case one has to manually inspect the integrand and choose the integration regions accordingly. A future version of TKWANT should certainly have an adaptative energy/momentum integration. This somewhat complicates the parallel implementation, however, as whenever a region is subdivided the wavefunctions need to be recomputed for the new energies/momenta. The question is then how to best divide the available computing resources between the various sub-regions.

4.3   TKWANT Usage in the Field: A Gallery of Examples

In this section we will see a small gallery of examples from work done primarily by my collaborators, which will complement the applications from my own work that will be seen in part II. While the interface of the current version of TKWANT is somewhat rough around the edges, the results presented in this section show nevertheless that it is a general tool that is capable of bringing real added value.

4.3.1   Calculating Time-Resolved Shot Noise

Single particle observables such as the charge and current can generally be written in terms of a single-particle Green's function. In the wavefunction formalism, these can be written as a single integral over energy, as seen in Sect. 3.1. Often, one wants to go beyond such simple observables and instead look at higher-order cumulants. Current-current correlations, for example, can be written as

$$\hat{S}_{\mu\nu}(t, t') = \left(\hat{I}_\mu(t) - \langle\hat{I}_\mu(t)\rangle\right) \times \left(\hat{I}_\nu(t') - \langle\hat{I}_\nu(t')\rangle\right) \tag{4.3}$$

where $\hat{I}_\mu(t)$ is the current operator for current flowing across an interface μ:

$$\hat{I}_\mu(t) = \sum_{(i,j)\epsilon\mu} \mathbf{H}_{ij}(t)\hat{c}_i^\dagger(t)\hat{c}_j(t) - \mathbf{H}_{ji}(t)\hat{c}_j^\dagger(t)\hat{c}_i(t), \tag{4.4}$$

and $\left\langle\hat{I}_\mu(t)\right\rangle$ is its thermal average. We can see that $\hat{S}_{\mu\nu}(t, t')$ will contain products of four operators and hence, using Wick's theorem, its thermal average can be expressed as products of *two* Green's functions.

In Ref. [19] expressions are derived for the current-current correlations in terms of the time-evolved scattering wavefunctions. Specifically, the thermal average of the current-current correlations can be written as

$$\langle\hat{S}_{\mu\nu}(t, t')\rangle = \sum_{\alpha\beta} \int \frac{dE}{2\pi} \int \frac{dE'}{2\pi} f_\alpha(E)[1 - f_\beta(E')]I_{\mu,EE'}(t)[I_{\nu,EE'}(t')]^* \tag{4.5}$$

where

$$I_{\mu,EE'}(t) = \sum_{(i,j)\epsilon\mu} \left(\mathbf{H}_{ij}(t)[\psi^\dagger_{\beta E'}(t)]_i[\psi_{\alpha E}(t)]_j - \mathbf{H}_{ji}(t)[\psi^\dagger_{\beta E'}(t)]_j[\psi_{\alpha E}(t)]_i \right).$$

(4.6)

In Ref. [19] they did not numerically compute such quantities directly, however, opting instead to look at the variance of the particle number operator:

$$\hat{n}_\mu = \int_{-\tau/2}^{\tau/2} dt\, \hat{I}_\mu(t),$$

(4.7)

which they split into three contributions in order to take care of terms divergent in the limit $\tau \to \infty$, which come from the contribution from the equilibrium. Specifically,

$$\text{var}\left(\hat{n}_\mu\right) \equiv \langle \hat{n}_\mu^2 - \langle \hat{n}_\mu \rangle^2 \rangle = \tau \sigma_{\text{st}}^2 + 2\sigma_{\text{mix}} + \bar{\sigma}^2 + \mathcal{O}\left(\frac{1}{\tau}\right)$$

(4.8)

where the three contributions are defined as

$$\sigma_{st}^2 = \sum_{\alpha\beta} \int \frac{dE}{2\pi} f_\alpha(E)[1 - f_\beta(E)] \left| I_{\mu,EE}(0) \right|^2$$

(4.9)

$$\sigma_{mix} = \sum_{\alpha\beta} \int \frac{dE}{2\pi} f_\alpha(E)[1 - f_\beta(E)] \Re \bar{N}^*_{EE} I_{\mu,EE}(0)$$

(4.10)

$$\bar{\sigma}^2 = \sum_{\alpha\beta} \int \frac{dE}{2\pi} \int \frac{dE'}{2\pi} f_\alpha(E)[1 - f_\beta(E')] \left| \bar{N}_{EE'} \right|^2$$

(4.11)

where

$$\bar{N}_{EE'} = \int_{-\infty}^{\infty} dt[I_{\mu,EE'}(t) - I_{\mu,EE'}(0)e^{-i(E-E')t/\hbar}]$$

(4.12)

They then use the raw output of TKWANT (the time-evolved scattering states) to calculate var $\left(\hat{n}_\mu\right)$ for a one-dimensional conductor with an insulating barrier (shown in Fig. 4.2a), when a voltage pulse is applied to one of the contacts. In addition they derived an analytical result for var $\left(\hat{n}_\mu\right)$ in this specific case and compared it to simulation; the results are shown in Fig. 4.2b, when the area underneath the voltage pulse $\bar{n}$ (= $(e/h) \int V(t)dt$) is varied. They saw that var $\left(\hat{n}_\mu\right)$ is *minimised* for integer values of $\bar{n}$, in analogy to the Levitons discussed in the introduction [20].

Although this result is in itself intriguing, it also paves the way for other time-resolved noise properties to be calculated with TKWANT. We note, however, that the problem required some delicate refactoring in order to avoid problems with infinities related to the equilibrium contributions, which may hinder attempts to make a robust implementation in the general case. The fact that there are now *two* energy integrals to calculate also somewhat complicates matters. However, the fact

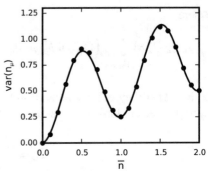

(a) Sketch of the one-dimensional wire system with insulating barrier used for calculating the noise in the number of transmitted particles. A gaussian voltage pulse $V(t)$ is applied to the left-hand contact, which injects charge into the system. The insulating barrier, controlled with V_g, provides backscattering. The number of particles traversing the cross-section μ is measured, as is its noise properties.

(b) The variance in the number of particles n_μ traversing the cross-section μ versus the number of injected particles $\bar{n}$, computed from TK-WANT simulations using eqs. (4.8) and (4.9).

Fig. 4.2 Computing the noise in a quantum wire after the application of a voltage pulse. Both subfigures reproduced with permission from Ref. [19]

that the two-energy quantity $I_{\mu,EE'}(t)$ can be expressed in terms of just the single-energy quantity $\psi_{\alpha E}(t)$ indicates that the computational effort will still scale linearly as a function of the number of energy points. This is because computing $\psi_{\alpha E}(t)$ on a set of energies $\lambda = \{E_0, E_1, \ldots, E_N\}$ allows us to directly evaluate $I_{\mu,EE'}(t)$ on the *grid* $\lambda \times \lambda = \{(E_0, E_0), (E_0, E_1), \ldots, (E_N, E_{N-1}), (E_N, E_N)\}$. Whether the density of points required to reach convergence will be higher when calculating the noise compared to the current is not clear.

4.3.2   Stopping Voltage Pulses in the Quantum Hall Regime

The integer quantum Hall effect (IQHE) is observed in two-dimensional systems subject to strong magnetic fields [21], characterised by an insulating bulk and unidirectional edge states [22]. While most applications of the IQHE concentrate on the properties of these edge states, in Ref. [23] TKWANT is used to investigate the crossover between these edge states and the bulk insulating states. Figure 4.3 shows a sketch of the simulated setup: a two-dimensional system in the quantum Hall regime (light grey) is connected to two electrodes (black rectangles) and a gate with potential $V_g(t)$ is electrostatically coupled to the system (dashed box). A voltage pulse is applied to the lower contact, which injects charge into the system (the dark blob moving through the system in Fig. 4.3. Initially the gate is at a potential V_0, which means that the right-hand "edge" of the system is found in the centre, and we see

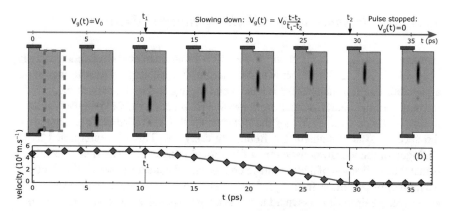

Fig. 4.3 *Upper* Snapshots of the deviation from equilibrium of the charge density as a function of time in a two-dimensional system in the quantum Hall regime. The *dark blob* is the charge pulse discussed in the main text. The system (*light grey*) is connected to two electrodes (*black rectangles*) and a gate voltage is applied on *top* of the *right-hand portion* of the system (*dashed box*). The gate voltage is initially V_0, and is reduced to 0 between t_1 and t_2. *Lower* Velocity of the charge pulse as a function of time. TKWANT is used to calculate $\langle \hat{y}(t) \rangle$ and this is differentiated numerically. Reproduced with permission from Ref. [23]

that the charge pulse propagates along this edge channel. Afterwards, at a time t_1, the gate voltage is lowered until time t_2 when it becomes 0. This effectively shifts the right-hand edge channels to the true edge of the system. We can see from Fig. 4.3, however, that the charge pulse does *not* follow the evolution of the edge channel; instead, it stays in the centre of the system and its velocity decreases. We see that after t_2 the pulse has completely stopped in the centre of the system. This behaviour is independent of the time $t_2 - t_1$ over which the gate voltage is lowered, provided that this time is not so long as to let the charge pulse escape through the upper contact. The fundamental reason for this highly non-intuitive result is that the system remains translationally invariant along the y direction at all times, which means that the quasimomentum k must be conserved (a more in-depth explanation is given in Refs. [23, 24]).

Intuitively one would perhaps think that the charge pulse would follow the edge channel as the gate voltage is decreased, however the simulations indicate that this intuition is incorrect. This is a prime example of where numerics in general (and TKWANT specifically) offer a great deal of added value. Without the initial insight offered by the TKWANT simulations, it would initially not have been clear that there was any interesting physics to be seen in the above-described setup. While a deeper understanding will usually be found by employing analytical methods, the easy-to-perform numerics afforded by TKWANT offer increased agility and ability to concentrate on the "big picture", which is invaluable in the initial stages of an exploratory research project.

4.3.3   a.c. Josephson Effect Without Superconductivity

In many of the applications in part II we will see that quantum interference will play
a big role. In Ref. [25] it was shown that after an abrupt change of bias voltage across
a quantum interferometer there is a universal regime where the current measured at
the output of the interferometer oscillates at a frequency $eV_b/\hbar$ (V_b the bias voltage)
during a certain time τ_f. While the duration of this universal regime depends on
the specific system studied, the frequency does not. This was interpreted as the
analogue of the a.c. Josephson effect for normal conductors. The fundamental idea
is that the application of a bias voltage changes the frequency at which the electronic
wavefunctions oscillate, from $E/\hbar$ to $(E + eV_b)/\hbar$. This frequency change originates
at the voltage drop (it is assumed that this drop is spatially short, and occurs before
the actual interferometer) and propagates through the system at the electronic group
velocity. As the interferometer will naturally have multiple paths of different lengths,
the frequency change will take different times to propagate along each path. Once
the frequency change has arrived from the shortest path, the part of the wavefunction
oscillating at $(E + eV_b)/\hbar$ will interfere with the part oscillating at $E/\hbar$, giving rise
to oscillations of frequency $eV_b/\hbar$. In Ref. [25] a Mach-Zehnder interferometer in
the quantum Hall regime is studied due to its simplicity (only two paths are available
through the system due to the quantised edge channels); this is illustrated in Fig. 4.4a.
Figure 4.4b shows the current measured at contact 1 as a function of time after the bias
on contact 0 is raised to V_b at $t = 0$. We clearly see that the output current oscillates
at frequency $eV_b/\hbar$ for a finite time before we reach the steady state regime that one
would expect from a d.c. bias.

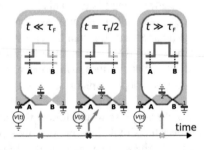

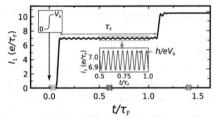

(a) Snapshots of the charge density (measured
from equilibrium) in the Mach-Zehnder interferometer at different times after the bias on
contact 0 is raised, from vanishing density to
$10^{11}\,\mathrm{cm}^{-2}$.

(b) Current at contact 1 as a function of time. *Upper inset*: Schematic of the voltage rise. *Lower
inset*: zoom of the transient regime. The
crosses on the time axis correspond to the
times at which the snapshots in fig. 4.4a are
taken.

Fig. 4.4 Results from TKWANT simulations of the Mach-Zehnder interferometer after an abrupt
raise of bias voltage on contact 0. Subfigures reproduced (modified) from Ref. [25] under the
conditions of the Creative Commons CC-BY license

The strength of TKWANT in this work was that it was easy to set up simulations for other types of interferometers (a Fabry-Perot was also studied in Ref. [25]), which enabled the claim of a "universal" transient regime to be corroborated.

4.3.4 Floquet Topological Insulators

Systems with so-called "topological" phases have recently seen a great deal of theoretical as well as experimental interest [26]. Such systems are characterised by a bulk that is insulating and surfaces that support chiral edge states, and are thus are conducting. One simple example of this is the quantum Hall regime that can be induced in two-dimensional electron gases by applying a strong perpendicular magnetic field [21, 27]. A more recent development involves inducing topological phases by applying *periodic perturbations* to the system; these are known as Floquet topological insulators [28–30].

In Ref. [31] TKWANT is used to show that there is a correspondence between the differential conductance and the quasienergy spectrum of a Floquet topological insulator that arises from the Bernevig-Hughes-Zhang (BHZ) model [32, 33], which is used as a model for mercury telluride/cadmium telluride quantum wells [33, 34]. Schematically, the half-BHZ model corresponds to a square lattice model with two orbitals per site and first and second nearest neighbour hoppings:

$$\hat{\mathbf{H}} = \sum_{x,y} [(A\sigma_3 - B\sigma_0)|x, y\rangle\langle x, y| + C\sigma_0|x, y\rangle\langle x + 1, y| + D\sigma_0|x, y\rangle\langle x, y + 1|+$$

$$J\sigma_0|x, y\rangle\langle x + 1, y + 1| + |x, y\rangle\langle x + 1, y - 1|)], \tag{4.13}$$

where capital latin letters denote constant scalars that parametrise the model, and σ_j are Pauli matrices:

$$\sigma_0 = \begin{pmatrix} 1 & 0 \\ 0 & 1 \end{pmatrix}, \sigma_1 = \begin{pmatrix} 0 & 1 \\ 1 & 0 \end{pmatrix}, \sigma_2 = \begin{pmatrix} 0 & -i \\ i & 0 \end{pmatrix}, \sigma_3 = \begin{pmatrix} 1 & 0 \\ 0 & -1 \end{pmatrix}, \tag{4.14}$$

Additionally, a periodic on-site perturbation

$$\Delta\mathbf{H}_{\text{BHZ}} = \sum_{x,y} F[\sin(\omega t)\sigma_1 + \cos(\omega t)\sigma_2] \tag{4.15}$$

is added that can induce a topological phase. As the Hamiltonian is time dependent energy is no longer conserved, however—as the perturbation is periodic with period T—the *quasienergy* is still a useful quantity. The quasienergies ε_α are defined by the eigenvalues of the *Floquet operator* $\hat{U}(T, 0)$:

$$\hat{U}(T, 0)|\varphi_\alpha\rangle = e^{-i(\varepsilon_\alpha - i\eta_\alpha)T/\hbar}|\varphi_\alpha\rangle \tag{4.16}$$

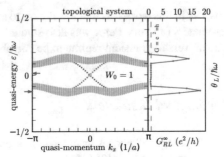

(a) Quasienergy spectrum (left) and the differential conductance (right) for the two-terminal setup when the model is tuned to exhibit a Floquet topological phase. The symbols in the gap between the two quasi-energy bands show the topological Floquet (edge) states. The differential conductance shows finite quantisation e^2/h in the quasienergy gap.

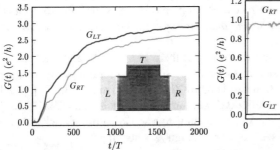

(b) Conductance as a function of time for the three-terminal setup, at the Fermi energy indicated by the arrow near 0 quasienergy in fig. 4.5a. *Inset*: probability density map at $t/T = 1800$, showing the delocalised Floquet state.

(c) Conductance as a function of time for the three-terminal setup, at the Fermi energy indicated by the arrow at negative quasienergy in fig. 4.5a. *Inset*: probability density map at $t/T = 1800$, showing the Floquet state localised on one edge.

Fig. 4.5 Results from TKWANT simulations of the half-BHZ model (see main text). Subfigures reproduced with permission from Ref. [31]

where $\eta_\alpha/\hbar$ is a damping rate, and $|\psi_\alpha(t)\rangle$ is a right eigenstate of the Floquet operator. The Floquet operator itself is just the evolution operator evaluated over one period:

$$\hat{U}(T, 0) = \mathcal{T}\left[e^{-(i/\hbar)\int_0^T \hat{\mathbf{H}}(t)dt}\right], \qquad (4.17)$$

where $\mathcal{T}$ is the time-ordering operator.

First the case of a two-terminal quasi one-dimensional system is analysed. The model is placed in a parameter regime that should exhibit topological states and the quasienergy spectrum is calculated analytically, while TKWANT is used to calculate the time average of the differential conductance. Figure 4.5a shows this quasienergy spectrum and differential conductance. The differential conductance clearly shows two peaks where the two bands of quasienergies occur, and the conductance drops to a *finite* value e^2/h in the "gap" between the bands. This finite conductance is due to the presence of topological edge states inside the gap (shown as the symbols in gap in the quasienergy spectrum).

This can be seen even more clearly by performing a three-terminal measurement in a T-junction geometry. Figure 4.5b and c show the three-terminal differential conductance as a function of time after the perturbation is switched on at two different chemical potentials. Figure 4.5b corresponds to a chemical potential inside the band of quasienergies, whereas Fig. 4.5 corresponds to a chemical potential in the gap (these are marked by arrows on the y-axis in Fig. 4.5a). In the former case we clearly see that the both the conductances take unquantised, finite values and the inset of Fig. 4.5b shows the delocalised nature of the Floquet state. In the latter case, however, we clearly see that the conductance between the left and top contacts remains zero while the between the right and top contacts the conductance is finite and (nearly) quantised to e^2/h. In addition the inset of Fig. 4.5c clearly shows the localised nature of the Floquet state inside the gap.

The use of TKWANT in this work clearly demonstrates its applicability to purely periodic problems, despite the fact that it is a time-resolved approach.

References

1. G. Van Rossum, *Python Programming Language*, https://www.python.org/
2. Y. Matsumoto, *Ruby Programming Language*, https://www.ruby-lang.org/en/
3. List of quantum chemistry and solid-state physics software. Page Version ID: 716155842. April 2016, https://en.wikipedia.org/w/index.php?title=List_of_quantum_chemistry_and_solid-state_physics_software
4. Cyril Labbe and Dominique Labbe, Scientometrics Duplicate and fake publications in the scientific literature: how many SCIgen papers in computer science? Scientiometrics **94**(1), 379–396 (2012)
5. R. Van Noorden, Publishers withdraw more than 120 gibberish papers. Nature (2014)
6. L. Susan Blackford et al., An updated set of basic linear algebra subprograms (BLAS). ACM Trans. Math. Softw. **28**(2), 135–151 (2002)
7. E. Anderson (ed.), LAPACK users' guide, 3rd edn, in *Software, Environments, Tools* (Society for Industrial and Applied Mathematics, Philadelphia, 1999)
8. R. Piessens et al., Quadpack, vol. 1. *Springer Series in Computational Mathematics* (Springer, Berlin, 1983)
9. What is Free Software? https://www.gnu.org/philosophy/free-sw.html
10. NextNano GmbH. NextNano++, http://www.nextnano.com/nextnanoplus/
11. Quantumwise A/S. Atomistix Toolkit, https://www.quantumwise.com
12. NemoCo. Nemo 5, https://engineering.purdue.edu/gekcogrp/software-projects/nemo5/
13. Python Package Index, https://pypi.python.org/pypi
14. S. van der Walt, S.C. Colbert, G. Varoquaux, The NumPy array: a structure for efficient numerical computation. Comput. Sci. Eng. **13**(2), 22–30 (2011)
15. SymPy Development Team, *SymPy: Python Library for Symbolic Mathematics* (2016), http://www.sympy.org
16. J.D. Hunter, Matplotlib: a 2D graphics environment. Comput. Sci. Eng. **9**(3), 90–95 (2007)
17. S. Behnel et al., Cython: the best of both worlds. Comput. Sci. Eng. **13**(2), 31–39 (2011)
18. C.W. Groth et al., New J. Phys. **16**(6), 063065 (2014)
19. B. Gaury, X. Waintal, A computational approach to quantum noise in time-dependent nano-electronic devices. Phys. E: Low-Dimension. Syst. Nanostruct. **75**, 72–76 (2016)

20. J. Dubois et al., Integer and fractional charge Lorentzian voltage pulses analyzed in the framework of photon-assisted shot noise. Phys. Rev. B **88**(8) (2013)

21. K.V. Klitzing, G. Dorda, M. Pepper, New method for high-accuracy determination of the fine-structure constant based on quantized hall resistance. Phys. Rev. Lett. **45**(6), 494–497 (1980)

22. M. Büttiker, Absence of backscattering in the quantum Hall effect in multiprobe conductors. Phys. Rev. B **38**(14), 9375–9389 (1988)

23. B. Gaury, J. Weston, X. Waintal, Stopping electrons with radio-frequency pulses in the quantum Hall regime. Phys. Rev. B **90**(16), 161305 (2014)

24. B. Gaury, Emerging concepts in time-resolved quantum nanoelectronics. Ph.D. thesis. Université de Grenoble, October 2014

25. B. Gaury, J. Weston, X. Waintal, The a.c. Josephson effect without superconductivity. Nat. Commun. **6**, 6524 (2015)

26. M.Z. Hasan, C.L. Kane, Colloquium: topological insulators. Rev. Mod. Phys. **82**(4), 3045–3067 (2010)

27. D.J. Thouless et al., Quantized Hall conductance in a two-dimensional periodic potential. Phys. Rev. Lett. **49**(6), 405–408 (1982)

28. T. Kitagawa et al., Topological characterization of periodically driven quantum systems. Phys. Rev. B **82**(23), 235114 (2010)

29. J.P. Dahlhaus et al., Quantum Hall effect in a one-dimensional dynamical system. Phys. Rev. B **84**(11), 115133 (2011)

30. J. Cayssol et al., Floquet topological insulators. Phys. Status Solidi RRL **7**(1–2), 101–108 (2013)

31. M. Fruchart et al., Probing (topological) floquet states through DC transport. Phys. E: Low-Dimension. Syst. Nanostruct. **75**, 287–294 (2016)

32. M.S. Rudner et al., Anomalous edge states and the bulk-edge correspondence for periodically driven two-dimensional systems. Phys. Rev. X **3**(3), 031005 (2013)

33. B.A. Bernevig, T.L. Hughes, S.-C. Zhang, Quantum spin Hall effect and topological phase transition in HgTe quantum wells. Science **314**(5806), 1757–1761 (2006)

34. M. König et al., Quantum spin Hall insulator state in HgTe quantum wells. Science **318**(5851), 766–770 (2007)

Part II
Applications of the Numerical Algorithms

Chapter 5
Split Wire Flying Qubit

In recent years there has been a big push to develop platforms for quantum computation. Many of the common proposals encode the quantum information (or *qubits*) in localised states (e.g. on a quantum dot [1, 2], or in a superconducting junction [3–6]). An alternative proposal, referred to as a "flying qubit" consists of encoding the information in a state with a finite velocity [7], so that the qubits can have gate operations applied *in flight* as they are moved around to different parts of the circuit. Recently a split wire geometry proposed to host and manipulate these flying qubits was realised experimentally, and has since seen an increased experimental as well as theoretical interest [8–11]. Previous efforts to simulate this system were limited to looking at d.c. physics [10], or did not take into account the Fermi statistics of the electrons in the system [7]. Experiments currently being carried out in the group of Christopher Baüerle at the Néel Institute in Grenoble hope to directly probe the time-resolved behaviour of such a split-wire flying qubit. With the source-sink algorithm we are perfectly positioned to numerically investigate the behaviour of these flying qubits in the time domain, which will aid in interpretation of experimental data and design of future generations of devices. In this chapter we will start by recovering previously obtained results for the d.c. behaviour of the split wire system, in order to illustrate the main physical effect at play: quantum interference between different paths through the system. This will be followed by *novel time-resolved simulations* when a voltage pulse is applied to one of the electrodes of the split wire. We will see the emergence of a *dynamical modification* of the interference compared to the d.c. case, a concept that was recently elucidated in a number of publications [12, 13].

5.1 Simple Model for the Split Wire Geometry

Figure 5.1a shows the experimental setup that we are going to model. The setup consists of a gallium arsenide-aluminium gallium arsenide heterostructure (dark background) with metallic gates (the light grey shapes) deposited on the surface.

© Springer International Publishing AG 2017

J. Weston, *Numerical Methods for Time-Resolved Quantum Nanoelectronics*,
Springer Theses, DOI 10.1007/978-3-319-63691-7_5

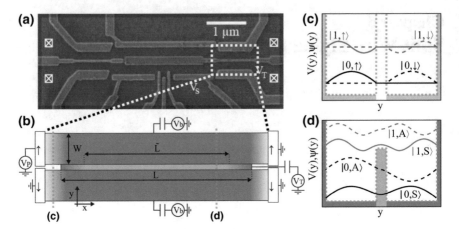

Fig. 5.1 Split wire setup and transverse modes. **a** Scanning electron microscope image of the experimental setup (reprinted with permission from Ref. [10] Copyright 2015 by the American Physical Society). **b** Sketch of our model setup, the gate voltage V_T controls the tunnel barrier, V_b controls the charge density in the coupling region, and V_p applied to the *upper-left* lead can inject charge into the system. L measures the total coupling region length, $\tilde{L}$ measures the effective coupling region length, and W measures the width of an individual wire. The *dashed lines* labelled (**c**) and (**d**) refer to the cross sections shown in the remaining subfigures. **c** Sketch of the 4 lowest energy transverse modes before the coupled region. along with the transverse potential, $V(y)$. The states $|n, \uparrow\rangle$ and $|n, \downarrow\rangle$ are degenerate. **d** Sketch of the 4 lowest energy transverse modes in the coupled region, along with the transverse potential, $V(y)$

A two-dimensional electron gas (2DEG) forms at the interface between the two materials (parallel to the page), and ohmic contacts (on the extreme right/left of the sample, out of the view of Fig. 5.1a) allow for charge to be injected into the 2DEG. Voltages applied to the metallic gates allow the charge to be confined to restricted regions of the device. We shall be interested in the region of the device outlined in Fig. 5.1a, which we model as two quasi one-dimensional wires of width W that are coupled in some finite region; this is shown in Fig. 5.1b. There is a potential barrier between the two wires controlled by the parameter V_T; this controls the tunnelling between the two wires. While the wires in our model are formally coupled over a length L, we actually reduce the potential adiabatically along the x direction from a large value V_∞ to V_T (the purple colour gradient in Fig. 5.1b) in order to min-imise the reflection from the boundary between the coupled/uncoupled regions. The consequence of this is that the *effective* length over which the wires are coupled is smaller, $\tilde{L}$. In addition we model the potential provided by the side-gates (light grey in Fig. 5.1a) as a hard-wall boundary in the y direction, in addition to a uniform potential in the coupled region. This potential is uniform over the coupled region, controlled by V_b, and falls adiabatically to zero outside the coupled region (the grey colour gradient in Fig. 5.1b); this allows us to control the number of open conduction channels in the coupled region. We ground three of the contacts and apply a (possibly time-dependent) voltage V_p to lead $\uparrow$ on the left. We model the voltage drop as being abrupt between the contact and the scattering region.

As the Fermi wavelength is sufficiently long compared to the inter-atomic distance of the material ($\lambda_F \approx 44$ nm for the electron density used in the sample of Ref. [10]) we can model the system with a continuum Hamiltonian:

$$\hat{H}(t) = \int_{-\infty}^{\infty} dx \int_{-W}^{W} dy \hat{\psi}^{\dagger}(x, y) \left[-\frac{\hbar^2}{2m^*} \nabla^2 + qU_T(x)\delta(y) + qU_b(x) \right.$$
$$\left. + qU_p(t)\Theta(x - M)\Theta(-y) \right] \hat{\psi}(x, y), \tag{5.1}$$

where m^* is the effective mass of the 2DEG, q is the electronic charge, $M = L/2 + 5\chi$ (χ is a length that determines the scale of the variations in $U_b(x)$ and $U_T(x)$) and $\Theta(y)$ and $\delta(y)$ are Heaviside functions and Dirac delta functions respectively. The $\hat{\psi}^{\dagger}(x, y)$ ($\hat{\psi}(x, y)$) are creation (destruction) operators for single particle states at position (x, y). The potentials take the following form:

$$U_T(x) = V_T + V_{\infty} \left\{ 1 - \frac{1}{2} \left[\tanh\left(\frac{x + L/2 - 5\chi}{\chi}\right) + \tanh\left(\frac{x - L/2 + 5\chi}{\chi}\right) \right] \right\} \tag{5.2}$$

$$U_b(x) = \frac{V_b}{2} \left[\tanh\left(\frac{x + L/2 + 5\chi}{\chi}\right) - \tanh\left(\frac{x - L/2 - 5\chi}{\chi}\right) \right], \tag{5.3}$$

where V_{∞} is a value sufficiently large so as to render the leads effectively uncoupled just at the start of the coupled region (to enable a smooth transition between these regions). We discretise this Hamiltonian onto a square lattice of spacing a, using the procedure outlined in appendix A, and perform a gauge transformation to bring the time-dependence from the lead $\uparrow$ on the left into the coupling between the lead and the scattering region (see appendix C for details). In what follows we will express all energies in units of the tight-binding bandwidth $\gamma_B = 4\hbar^2/(2m^*a^2)$, voltages in units of γ_B/e and times in units of $\hbar/\gamma_B$.

5.2 d.c Characterization of the Split Wire

In this section we will characterise the d.c. behaviour of the device. We shall see that the split wire can be considered as an effective two-path interferometer; this point of view will be invaluable when interpreting the time-resolved simulations in Sect. 5.3. In order to calculate the d.c. conductance $G_{\sigma'\sigma}$ between lead σ on the left and lead σ' on the right ($\sigma \in \{\uparrow, \downarrow\}$) we need only use the Landauer formula [14, 15]:

$$G_{\sigma'\sigma} = \frac{2e^2}{h} D_{\sigma'\sigma} \tag{5.4}$$

where $D_{\sigma'\sigma}$ is the transmission from lead σ on the left to lead σ' on the right, defined by

$$D_{\sigma'\sigma} = \sum_{n,m} T_{m\sigma',n\sigma} \tag{5.5}$$

where $T_{m\sigma',n\sigma}$ is the transmission *probability* from mode $|n, \sigma\rangle$ on the left to mode $|m, \sigma'\rangle$ on the right (these modes are sketched in Fig. 5.1c). In all that follows we shall assume that inter-band scattering is negligible, i.e. $T_{m\sigma',n\sigma} = \delta_{mn} T_{n\sigma',n\sigma}$ where δ_{mn} is the Kronecker delta.

5.2.1   Analytical Treatment Using Scattering Theory

The $T_{n\sigma',n\sigma}$ can be calculated by using a wave matching procedure; here we will reproduce previously perfomed calculations [10] to highlight the salient physics of the system, while avoiding the intricacies of the more complicated model presented in Sect. 5.1. In Sect. 5.2.2 we will treat the model numerically, which will allow us to validate this simplified analytical treatment.

The full wavefunction in the uncoupled region can be written $\Psi_{n,\sigma}(x, y) = \langle y|n, \sigma\rangle e^{ik_{n,\sigma}x}$, where $\sigma \in \{\uparrow, \downarrow\}$, and its energy is $E = E_{n,\sigma} + (\hbar^2/2m^*)k_{n,\sigma}^2$, where $E_{n,\sigma}$ is the energy of the transverse mode $|n, \sigma\rangle$, and $k_{n,\sigma}$ is the longitudinal wavevector. As the states $|n, \uparrow\rangle$ and $|n, \downarrow\rangle$ are degenerate for a given n, we can also define symmetric and antisymmetric superpositions:

$$|n, \uparrow\rangle = \frac{1}{\sqrt{2}} [\,|n, S_u\rangle + |n, A_u\rangle]$$
$$|n, \downarrow\rangle = \frac{1}{\sqrt{2}} [\,|n, S_u\rangle - |n, A_u\rangle]\,, \tag{5.6}$$

where the u subscript reminds us that these are transverse modes in the *uncoupled* region. In the coupled region we also have symmetric and antisymmetric modes $|n, S\rangle$ and $|n, A\rangle$ (illustrated in Fig. 5.1d), and we suppose that the transition from the uncoupled to the coupled region is adiabatic, such that $|n, S_u\rangle$ evolves into $|n, S\rangle$ and $|n, A_u\rangle$ evolves into $|n, A\rangle$ with no inter-mode scattering. While $|n, A_u\rangle$ and $|n, S_u\rangle$ are degenerate, $|n, A\rangle$ and $|n, S\rangle$ are not. This means that for a given energy the states will have different longitudinal wavevectors, $k_{n,A}$ and $k_{n,S}$. If we are in a state $|n, \uparrow\rangle$ in the uncoupled region on the left, this means that a length $\tilde{L}$ after the wires are coupled we will be in a state:

$$|\psi_{n,\uparrow}\rangle = \frac{1}{\sqrt{2}} \left[e^{ik_{n,A}\tilde{L}}|n, A\rangle + e^{ik_{n,S}\tilde{L}}|n, S\rangle \right]. \tag{5.7}$$

The wires are then adiabatically uncoupled (near the right-hand leads) and we can write the state as the following decomposition on the states $|n, \uparrow\rangle$ and $|n, \downarrow\rangle$ on the right:

$$|\psi_{n,\uparrow}\rangle = \frac{1}{2}\left[(e^{ik_{n,S}\tilde{L}} + e^{ik_{n,A}\tilde{L}})|n, \uparrow\rangle + (e^{ik_{n,S}\tilde{L}} - e^{ik_{n,A}\tilde{L}})|n, \downarrow\rangle\right]. \quad (5.8)$$

We immediately see that the difference in wavevectors will give rise to interference between the symmetric and antisymmetric components. We can thus write down the transmission *amplitudes* for arriving on the right in $|n, \uparrow\rangle$ or $|n, \downarrow\rangle$ given that we were injected on the left in $|n, \uparrow\rangle$:

$$t_{n\uparrow,n\uparrow} = \exp\left(i\frac{k_{n,A} + k_{n,S}}{2}\tilde{L}\right)\cos\left(\frac{\Delta k_n}{2}\tilde{L}\right)$$
$$t_{n\downarrow,n\uparrow} = i\exp\left(i\frac{k_{n,A} + k_{n,S}}{2}\tilde{L}\right)\sin\left(\frac{\Delta k_n}{2}\tilde{L}\right), \quad (5.9)$$

where $\Delta k_n = k_{n,S} - k_{n,A}$. The transmission probabilites can be calculated from these amplitudes using $T_{n\sigma',n\sigma} = \left|t_{n\sigma',n\sigma}\right|^2$.

5.2.2 Numerical Treatment

Having an intuitive picture of the physics at play in the system, we shall now numerically study the model presented in Sect. 5.1 using the KWANT [16] package. In addition to providing a visualisation of the concepts developed in Sect. 5.2.1 it will also allow us to verify that our model conforms to this simplified view.

Figure 5.2 shows how the wavevector difference changes as a function of the coupling gate voltage V_T and the effect that this has on the transmission $D_{\uparrow\uparrow}$ from lead $\uparrow$ on the left to lead $\uparrow$ on the right. We clearly see regular oscillations when the wavevector difference changes linearly. As we go to to very high gate voltages we effectively uncouple the two wires, which explains why $D_{\uparrow\uparrow} \rightarrow 1$ in this limit. The dashed line in Fig. 5.2a shows $D_{\uparrow\uparrow}$ calculated using Eq. (5.9) where Δk_0 has been calculated numerically from the tight binding model; we see a good fit between the model and the simple analytical result.

Figure 5.3 shows the dispersion relations for the leads (subfigure a) and in the coupled region (subfigure b) calculated from the tight-binding model. We see in Fig. 5.3c the transmission probabilities for being transmitted through the first and second modes from lead $\uparrow$ on the left to lead $\uparrow$ on the right. We see that the transmission probabilities are 0 before the corresponding modes in the central region open. Note that in order for $T_{n\sigma',n\sigma}$ to be different from 0 we need *both* modes $|n, A\rangle$ *and* $|n, S\rangle$ to be open in the coupled region, as $|n, \sigma\rangle$ is a linear combination of both. We see that the transmission probabilities oscillate as a function of energy. The reason for this is clear, as Fig. 5.3b clearly shows that Δk_n changes as a function of energy.

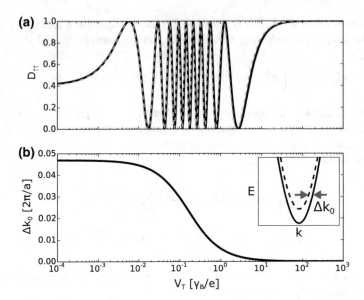

Fig. 5.2 d.c. simulation of split wire with $L = 700a$, $W = 10a$, $V_b = 0.11\gamma_B/e$ and $E_F = 0.15\gamma_B$. At this Fermi energy only the modes $|0, \uparrow\rangle$ and $|0, \downarrow\rangle$ in the coupled region are open. **a** *solid line*: transmission calculated from tight-binding simulation, *dashed line*: transmission calculated using Eq. (5.9) with Δk_0 calculated from tight-binding and $\tilde{L}$ as a fitting parameter. We used $\tilde{L} = 1242a$. **b** Tight-binding calculation of the difference in momentum between symmetric and antisymmetric modes in the coupling region. Both plots share the x-axis V_T scale

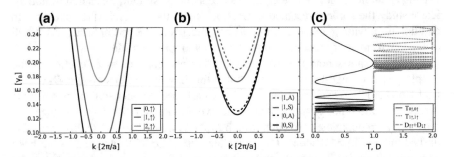

Fig. 5.3 Dispersion relations and transmissions for the split wire with $L = 700a$, $W = 10z$, $V_T = 0.27\gamma_B/e$ and $V_b = 0.11\gamma_B/e$. **a** Dispersion relation in lead $\uparrow$ showing the three lowest energy modes. **b** Dispersion relation in the coupling region of the split wire, showing the first 2 symmetric (*solid lines*) and anti-symmetric (*dashed lines*) modes. **c** Transmission probability from $|0, \uparrow\rangle$ on the *left* to $|0, \uparrow\rangle$ on the *right* (*black solid line*); transmission probability from $|1, \uparrow\rangle$ on the *left* to $|1, \uparrow\rangle$ on the *right* (*red dotted line*, shifted by 1 for clarity); total transmission from the $\uparrow$ lead on the *left* to the leads on the *right* (*green dashed line*). All three plots share the y-axis energy scale

The inter-band transmission probabilities $T_{m\sigma',n\sigma}$ (with $m \neq n$) are not shown, but are 0 at all energies (validating the assumptions of the analytical derivation above); this is because the transition from uncoupled to coupled region is done in an adiabatic manner.

One last point, which is perhaps a bit subtle, is that we expect to be able to see these interference effects even with a *large* number of open channels. Indeed, at the energy where the $n + 1$ channel opens the Δk_{n+1} is much larger than the Δk_n at the same energy (see Fig. 5.3b). This means that $T_{n\sigma',n\sigma}$ oscillates much slower than $T_{n+1\sigma',n+1\sigma}$ at the same energy, as can be clearly seen in Fig. 5.3c (at a given energy the red curve is oscillating much more quickly than the black curve). This separation in frequency of the different $T_{n\sigma',n\sigma}$ means that the oscillations from the different channels will be clearly distinguishable in the full differential conductance $G_{\sigma'\sigma}$.

5.3   Application of a Voltage Pulse

Now that we have the understanding of the system in d.c. we can now turn to time-resolved simulations. We apply a Gaussian voltage pulse to lead ↑ on the left and measure the current $I_\uparrow$ ($I_\downarrow$) leaving the system on the right via lead ↑ (↓). We also measure the current I_{in} injected into the system by the voltage pulse. We assume that the voltage drop is sharp and localised at the system-lead boundary. In addition, we tune V_b such that only the modes $|0, A\rangle$ and $|0, S\rangle$ are open in the coupled region at the Fermi energy. The pulses we apply have a full-width at half maximum of $200\hbar/\gamma_B$ and a typical height of $0.03\gamma_B/e$. The d.c. transmissions at the Fermi energy for the setup are $D_{\uparrow\uparrow} = 0.1$ and $D_{\downarrow\uparrow} = 0.9$. We have $L = 700a$ and $W = 10a$; in total we have 16700 sites in the scattering region (Fig. 5.4).

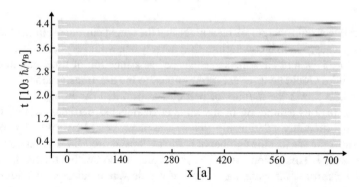

Fig. 5.4 Snapshots of the excess particle density in the split wire after the application of a voltage pulse that injects $\bar{n} = 0.1$ particles on average. Darker *grey* indicates positive deviations from the equilibrium particle density. The grey scale is independently normalised to the maximum density in each snapshot

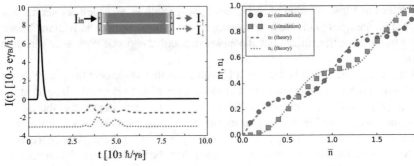

(a) Current as a function of time flowing in: lead ↑ on the left (I_{in}), lead ↑ on the right ($I_↑$), and lead ↓ on the right ($I_↓$), for a voltage pulse that injects $\bar{n} = 2.0$ particles.

(b) Number of transmitted particles on the right in lead ↑ ($n_↑$) and lead ↓ ($n_↓$) as a function of the injected number of particles ($\bar{n}$). Symbols: time-resolved simulation, dashed line: application of eq. (5.11) with $\tilde{L} = 596a$ as a fitting parameter.

Fig. 5.5 Charge transport after application of a voltage pulse on lead ↑ on the *left* of the split-wire. The system has $L = 700a$, $W = 10a$, $V_b = 0.11\gamma_B/e$, $V_T = 0.1446\gamma_B/e$ and $E_F = 0.15\gamma_B$. We use a pulse with a duration (full-width half-maximum) of $200\hbar/\gamma_B$. In d.c. the transmissions at the Fermi energy are $D_{\uparrow\uparrow} = 0.9$ and $D_{\downarrow\uparrow} = 0.1$

Figure 5.5a shows the results of a simulation where the above-defined currents are measured. Due to the large characteristic length $1/\Delta k_0$, and hence the large length of the system, we need to go to very long times (9500 times the inverse hopping parameter) in order to see the output current. The voltage pulse shown injects an average of $\bar{n} = 2$ particles into the system, where

$$e\bar{n} = \int_0^\infty dt \, I_{in}(t) \tag{5.10}$$

and e is the electronic charge. We clearly see that the output current oscillates between the ↑ and ↓ leads, which is counterintuitive; naïvely one would expect that the current in the two leads would have the same "shape" as a function of time, and that only the magnitudes would be different (proportional to the d.c. transmission).

Figure 5.5b shows the number of particles transmitted on the right into lead ↑ ($n_↑$) and ↓ ($n_↓$) as a function of the number of injected particles. Rather than a simple proportionality relationship (where the slope would be given by the d.c. transmission), we see that the number of particles depends non-linearly on $\bar{n}$ and even *oscillates* with $\bar{n}$. This curious behaviour can be understood within the framework of *dynamical control of interference*, aspects of which were elucidated in a number of recent publications [12, 13]. To understand this, one has to remember that there are already electrons in plane-wave states filling the Fermi sea before the voltage pulse is applied. The naïve picture that the electronic wavefunction will look like some sort of Gaussian wavepacket after the application of the voltage pulse is essentially wrong. Instead, the voltage pulse actually puts a *twist* in the phase of the plane-wave states

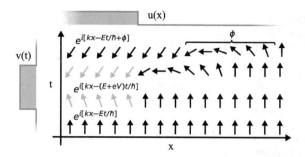

Fig. 5.6 Sketch of the effect of a voltage pulse $V(x, t) = v(t)u(x)$ on a plane-wave $\exp(ikx - iEt/\hbar)$. The $v(t)$ and $u(x)$ show the profile of the voltage pulse in time/space. The orientation of the arrows shows the phase of the wavefunction relative to a plane-wave $\exp(ikx - iEt)$. When the voltage pulse is applied, the wavefunction oscillates faster in time, and this phase "twist" φ propagates to the right

occupying the Fermi sea. To illustrate this, let us take a simple case where we have a perfect, infinite one-dimensional system with a time-dependent potential applied at $x < 0$. Let us take the case where the potential is initially zero and is abruptly raised to V at $t = 0$ and then lowered to 0 again at some later time t_1. Initially the scattering states originating from the left below the Fermi energy take the form of plane waves $\psi_{\alpha E}^{st}(t) = \exp(ik(E)x - iEt/\hbar)$. Just after the potential is raised the part of the wavefunction at $x < 0$ will now be oscillating *faster*, with frequency $(E + V)/\hbar$. As time passes, the part of the wavefunction that oscillates more quickly will propagate into the $x > 0$ region at a velocity $(1/\hbar)\partial E/\partial k$. When the potential is again lowered at $t = t_1$ the part of the wavefunction under the voltage pulse will again oscillate at frequency $E/\hbar$, however the part that propagated into the $x > 0$ region will still be oscillating at frequency $(E + V)/\hbar$. This is illustrated in Fig. 5.6. The phase of the wavefunction before and after the pulse are therefore offset by $(e/h)Vt_1$ with respect to one another (more generally they are offset by $\varphi = \int dt (e/h)V(t)$ for voltage pulses $V(t)$); the pulse induces a "phase domain wall" in the wavefunction. While this explanation is not rigorous, it can be shown that this intuitive picture is correct in the limit that the spectrum is linear (no dispersion) on energy scales of $\mathcal{O}(V)$ around the Fermi energy [17]. We can now employ this "phase domain wall" (PDW) picture to the present case to understand the source of the oscillations of $n_\uparrow$ ($n_\downarrow$) with $\bar{n}$. The voltage pulse creates a PDW that propagates into the split wire system. In the coupled region the state $|0, S\rangle$ has a larger velocity than $|0, A\rangle$ at a particular energy, which means that the PDW will travel faster along the antisymmetric component of the wavefunction than the symmetric one. If the pulse is sufficiently short with respect to $\Delta\tau_F = L/(v_{0,S} - v_{0,A})$ then the antisymmetric component will have its phase modified by φ at the output leads *before* the symmetric component, and the interference pattern will be modified during this finite interval, before returning to the d.c. interference pattern once the PDW has arrived from the (slower) symmetric component. Figure 5.7 illustrates this. The number of particles transmitted into each of the leads will therefore be affected by this modification of the interference pattern.

Fig. 5.7 Sketch of the propagation of the phase domain wall (PDW) along the two "paths" (symmetric $|0, S\rangle$ and antisymmetric $|0, A\rangle$ components) in the coupled region of the split wire. Between $t = L/v_{0,S}$ and $t = L/v_{0,A}$ the $|0, S\rangle$ component at the output on the right has had its phase modified, but the $|0, A\rangle$ component has not. The interference between the two components here will be different to that in d.c.; the interference has been modified *dynamically*

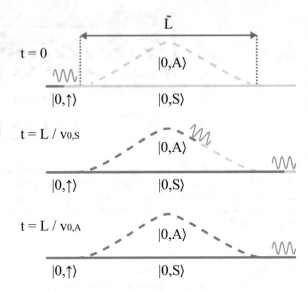

In Ref. [12] this reasoning is made more precise; in fact, as the split wire constitutes a two-path interferometer, the analysis for the present case is identical to that of the Mach-Zehnder interferometer studied in Ref. [12]. Concretely, we can apply Eqs. (27) and (28) of Ref. [12] to the split wire system and obtain the following relations:

$$n_\uparrow = \frac{\bar{n}}{2}\left[1 + \frac{1}{\pi}\sin(\pi\bar{n})\cos\left(\pi\bar{n} + \frac{\Delta k_0}{2}\tilde{L}\right)\right] \tag{5.11}$$

$$n_\downarrow = \frac{\bar{n}}{2}\left[1 - \frac{1}{\pi}\sin(\pi\bar{n})\cos\left(\pi\bar{n} + \frac{\Delta k_0}{2}\tilde{L}\right)\right]. \tag{5.12}$$

The lines in Fig. 5.5b correspond to the above analytical result where $\tilde{L}$ is used as a fitting parameter (we used $\tilde{L} = 596a$); we see a very good agreement with the numerics.

References

1. D. Loss, D.P. DiVincenzo, Quantum computation with quantum dots. Phys. Rev. A **57**(1), 120–126 (1998)
2. M. Veldhorst et al., A two-qubit logic gate in silicon. Nature **526**(7573), 410–414 (2015)
3. D. Vion et al., Manipulating the quantum state of an electrical circuit. Science **296**(5569), 886–889 (2002)
4. Yu. Chen et al., Qubit architecture with high coherence and fast tunable coupling. Phys. Rev. Lett. **113**(22), 220502 (2014)

5. Susanne Richer, David DiVincenzo, Circuit design implementing longitudinal coupling: a scalable scheme for superconducting qubits. Phys. Rev. B **93**(13), 134501 (2016)

6. R. Barends et al., Superconducting quantum circuits at the surface code threshold for fault tolerance. Nature **508**(7497), 500–503 (2014)

7. A. Bertoni et al., Quantum logic gates based on coherent electron transport in quantum wires. Phys. Rev. Lett. **84**(25), 5912–5915 (2000)

8. S. Takada et al., Measurement of the transmission phase of an electron in a quantum two-path interferometer. Appl. Phys. Lett. **107**(6), 063101 (2015)

9. S. Takada et al. Transmission phase in the kondo regime revealed in a two-path interferometer. Phys. Rev. Lett. **113**(12) (2014)

10. Tobias Bautze et al., Theoretical, numerical, and experimental study of a flying qubit electronic interferometer. Phys. Rev. B **89**(12), 125432 (2014)

11. Michihisa Yamamoto et al., Electrical control of a solid-state flying qubit. Nat. Nano **7**(4), 247–251 (2012)

12. B. Gaury, X. Waintal, Dynamical control of interference using voltage pulses in the quantum regime. Nat. Commun. **5**, 3844 (2014)

13. B. Gaury, J. Weston, X. Waintal. The a.c. Josephson effect without superconductivity. Nat. Commun. **6**, 6524 (2015)

14. R. Landauer, Spatial variation of currents and fields due to localized scatterers in metallic conduction. IBM J. Res. Dev. **1**(3), 223–231 (1957)

15. Rolf Landauer, Electrical resistance of disordered one-dimensional lattices. Philos. Mag. **21**(172), 863–867 (1970)

16. C.W. Groth et al., New J. Phys. **16**(6), 063065 (2014)

17. B. Gaury et al. Numerical simulations of time-resolved quantum electronics. Phys. Rep. **534**(1), 1–37 (2014)

Chapter 6
Time-Resolved Dynamics of Josephson Junctions

The source-sink algorithm presented in Sect. 3.3 allows us to perform time-resolved simulations of quantum transport that scale linearly with the required maximum simulation time. Such scaling is very desirable for systems with a large separation of energy (and hence time) scales. Systems that contain superconducting elements naturally have two energy scales, the Fermi energy (E_F) and the superconducting gap (Δ_0), that should be well separated if we wish to simulate systems in experimentally relevant regimes (for example $E_F/\Delta_0 \sim 10^{-5}$ for bulk aluminium at zero temperature). In this chapter we shall study the effect of time-dependent perturbations on Josephson junctions. We shall start with an introduction to the parts of the theory of superconductivity necessary to treat the problem with which we are concerned.[1] Having established the necessary background, we will focus on 3 physical effects. Firstly we will recover known physics of Josephson junctions: multiple Andreev reflection (MAR) and the a.c. Josephson effect. Secondly, we will discuss the *relaxation of the current* in a Josephson junction after an abrupt rise of the applied potential [1], showing how MAR comes into play in the relaxation rate. Finally, we will study an interesting phenomenon: the propagation of a voltage pulse through a Josephson junction [2]. We see that the pulse can become trapped in the junction, leading to a periodic current at the output that continues forever in the absence of additional relaxation mechanisms. After the completion of this work we became aware of previous work that follows a similar line to our own, but that used a different numerical technique [3].

---

[1]This introduction will stand us in good stead for Chap. 7 where we shall study more exotic superconducting states and Majorana quasiparticles.

© Springer International Publishing AG 2017
J. Weston, *Numerical Methods for Time-Resolved Quantum Nanoelectronics*,
Springer Theses, DOI 10.1007/978-3-319-63691-7_6

6.1   The Bogoliubov-de Gennes Equation

Conventional superconductivity is well described by the Bardeen-Cooper-Schrieffer (BCS) theory. Since their original proposal in 1957 [4, 5] there have been compatible reformulations of the theory that are more amenable for direct numerical simulation. Specifically, it turns out that all one needs to describe conventional superconductivity is to solve a Schrödinger-like equation in a Hilbert space twice the size of the one required for the problem without superconductivity. The "extended Schrödinger equation" to solve is is known as the *Bogoliubov-de Gennes* (BdG) equation [6], and has the following form in a discrete basis:

$$\begin{pmatrix} \mathbf{H} - E_F & \mathbf{\Delta} \\ \mathbf{\Delta}^* & E_F - \mathbf{H}^* \end{pmatrix} \begin{pmatrix} \mathbf{u}_n \\ \mathbf{v}_n \end{pmatrix} = E_n \begin{pmatrix} \mathbf{u}_n \\ \mathbf{v}_n \end{pmatrix}. \tag{6.1}$$

where $\mathbf{H}$ is the (possibly infinite) Hamiltonian matrix of the problem in the absence of superconductivity, and E_F is the Fermi energy. $\mathbf{\Delta}$ is a matrix defining the superconducting correlations (we shall define this more precisely in Sect. 6.1.2), $\mathbf{u}_n$ and $\mathbf{v}_n$ are vectors in the Hilbert space of the problem in the absence of superconductivity, and E_n is the energy. We can intuitively understand a few things about the BdG equation by considering the case where there are no superconducting correlations, $\mathbf{\Delta} = \mathbf{0}$. In this case we see that the BdG equations reduce to

$$\begin{aligned} (\mathbf{H} - E_F)\,\mathbf{u}_n &= E_n\,\mathbf{u}_n \\ (\mathbf{H} - E_F)\,\mathbf{v}_n^* &= -E_n\,\mathbf{v}_n^*, \end{aligned} \tag{6.2}$$

i.e. the two solutions correspond to solutions of the Schrödinger equation with energies E_n and $-E_n$ with respect to the Fermi energy; we thus identify the $\mathbf{u}_n$ and $\mathbf{v}_n$ parts of the solution as being associated with *electron* (above the Fermi energy) and *hole* (below the Fermi energy) excitations respectively. More generally we notice that if $(\mathbf{u}_n \ \ \mathbf{v}_n)^T$ is a solution to Eq. (6.1) with energy E_n, then $\left(-\mathbf{v}_n^* \ \ \mathbf{u}_n^*\right)^T$ is also a solution but with energy $-E_n$. Figure 6.1 shows the dispersion relation for an infinite 1D chain as we add the successive ingredients needed for superconductivity. We see the redundancy in the description using the BdG Hamiltonian, due to the "particle-hole" symmetry of the BdG Hamiltonian, exemplified by Fig. 6.1b; the same information is encoded in the electron (blue) states as the hole (red) states. We nevertheless require these two components; when $\mathbf{\Delta} \neq 0$, in Fig. 6.1c, the presence of the pairing term $\mathbf{\Delta} = \Delta_0 \mathbf{1}$ opens a gap of size $2\Delta_0$ in the spectrum around $E = 0$. Δ_0 is therefore referred to as the *superconducting gap*.

A Prescription for Time-Dependent Problems

As the BdG formalism deals with a single-particle Hamiltonian (albeit in a Hilbert space twice as large), we can use it in our framework for time-dependent transport. All we need do is to obtain the scattering states of the infinite system by solving Eq.

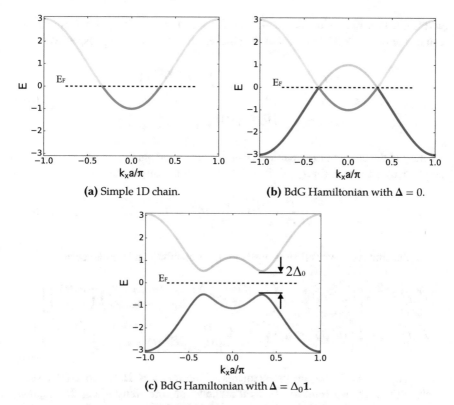

Fig. 6.1 Dispersion relations for a 1D chain with successive ingredients required for superconductivity. The line colouration depicts the electron-like (*blue*) or hole-like (*red*) nature of the states, *light/dark shading* depicts the states that are empty/filled at zero temperature. **a** and **b** describe the same physical situation (no superconductivity), and **c** describes a system with superconductivity

(6.1) instead of Eq. (3.2). We fill up these scattering states using the Fermi-Dirac distribution (noting that the Fermi level is now fixed at $E = 0$, due to the presence of the $-E_F$ in the definition of the BdG Hamiltonian), and evolve the scattering states using the generalisation of the time-dependent Schrödinger equation, Eq. (3.4):

$$i\hbar \frac{\partial}{\partial t} \begin{pmatrix} \mathbf{u}_{\alpha E}(t) \\ \mathbf{v}_{\alpha E}(t) \end{pmatrix} = \begin{pmatrix} \mathbf{H}(t) - E_F & \mathbf{\Delta}(t) \\ \mathbf{\Delta}^*(t) & E_F - \mathbf{H}^*(t) \end{pmatrix} \begin{pmatrix} \mathbf{u}_{\alpha E}(t) \\ \mathbf{v}_{\alpha E}(t) \end{pmatrix}. \qquad (6.3)$$

All that remains is to then calculate observables using Eq. (3.5) as usual.

6.1.1 Calculating Charge and Current in Superconducting Systems

Even though the prescription for calculating observables is clear in principle, one should nonetheless be careful when calculating densities and currents, as the wavefunction $(\mathbf{u} \ \mathbf{v})^T$ represents a state that *does not have a well-defined charge*. The **u** part is electron-like, with charge $-e$, and the $\mathbf{v}$ part is hole-like, with charge e. We can define the *probability* and *charge* densities on site i, ρ_i^p and ρ_i^c respectively, as

$$\rho_i^p(t) = \begin{pmatrix} u_i^*(t) & v_i^*(t) \end{pmatrix} \begin{pmatrix} 1 & 0 \\ 0 & 1 \end{pmatrix} \begin{pmatrix} u_i(t) \\ v_i(t) \end{pmatrix}$$

$$\rho_i^c(t) = q \begin{pmatrix} u_i^*(t) & v_i^*(t) \end{pmatrix} \begin{pmatrix} 1 & 0 \\ 0 & -1 \end{pmatrix} \begin{pmatrix} u_i(t) \\ v_i(t) \end{pmatrix}. \tag{6.4}$$

where q is the charge of the particles ($-e$ for electrons), and $u_i(t)$ and $v_i(t)$ are the components of $\mathbf{u}(t)$ and $\mathbf{v}(t)$ on site i. Using the continuity equation

$$\frac{\partial}{\partial_t} \rho_i^{p/c}(t) - \sum_j I_{ij}^{p/c} = 0, \tag{6.5}$$

where the sum runs over all sites, leads to the probability and charge currents:

$$I_{ij}^p(t) = \frac{2}{h} \Im \left[\begin{pmatrix} u_i^*(t) & v_i^*(t) \end{pmatrix} \begin{pmatrix} H_{ij}(t) - E_F \delta_{ij} & \Delta_{ij}(t) \\ \Delta_{ij}^*(t) & E_F \delta_{ij} - H_{ij}^*(t) \end{pmatrix} \begin{pmatrix} u_j(t) \\ v_j(t) \end{pmatrix} \right]$$

$$I_{ij}^c(t) = 2\frac{q}{h} \Im \left[\begin{pmatrix} u_i^*(t) & v_i^*(t) \end{pmatrix} \begin{pmatrix} H_{ij}(t) - E_F \delta_{ij} & \Delta_{ij}(t) \\ -\Delta_{ij}^*(t) & H_{ij}^*(t) - E_F \delta_{ij} \end{pmatrix} \begin{pmatrix} u_j(t) \\ v_j(t) \end{pmatrix} \right]. \tag{6.6}$$

where $H_{ij}(t)$ and $\Delta_{ij}(t)$ are the elements of the matrices $\mathbf{H}(t)$ and $\mathbf{\Delta}(t)$ respectively. Intuitively we think of I_{ij} as the "flow" of something along the hopping connecting sites i and j. Note that I^c is strange, however, as the component $I_{ii}^c(t) = 4(q/h)\Im\left[u_i^*(t)\Delta_{ii}(t)v_i(t)\right]$ is *non-zero* in the presence of superconductivity. What this means is that if we naively compute the sum of the currents flowing across the hoppings from site i to all connected sites j, the charge on site i will not be conserved due to this "onsite current" term I_{ii}^c.

6.1.2 From Second to First Quantisation

The preceding sections provide a complete scheme for calculating time-resolved observables in superconducting systems. For the sake of completeness we will now briefly review how Eq. (6.1) is obtained from the mean-field treatment of an interacting system. This is standard material and can be skipped by readers familiar with the theory of superconductivity.

We shall start from a Hamiltonian for electrons with a local, attractive two-body interaction,

$$\hat{H} = \sum_{ij\sigma}(h_{ij} - E_F\delta_{ij})\hat{c}_{i\sigma}^\dagger \hat{c}_{j\sigma} - V\sum_i \hat{c}_{i\uparrow}^\dagger \hat{c}_{i\downarrow}^\dagger \hat{c}_{i\downarrow}\hat{c}_{i\uparrow}, \tag{6.7}$$

where $\hat{c}^{\dagger}_{i\uparrow}$ ($\hat{c}_{j\downarrow}$) is an electron creation (destruction) operator for an electron with spin up (down) on site i (j), h_{ij} are the matrix elements for the non-interacting Hamiltonian, V is the strength of the interaction, and δ_{ij} is the Kronecker delta. Applying the mean field approximation to Eq. (6.7), following [6] we obtain

$$\hat{H}_{\text{eff}} = \sum_{ij} \left[\sum_{\sigma} \left(h_{ij} - E_F \delta_{ij} + U_{ij} \right) \hat{c}^{\dagger}_{i\sigma} \hat{c}_{j\sigma} + \Delta_{ij} \hat{c}^{\dagger}_{i\uparrow} \hat{c}^{\dagger}_{j\downarrow} + \Delta^*_{ij} \hat{c}_{j\downarrow} \hat{c}_{i\uparrow} \right] \quad (6.8)$$

with

$$U_{ij} = -V \langle \hat{c}^{\dagger}_{i\uparrow} \hat{c}_{i\uparrow} \rangle \delta_{ij}$$
$$\Delta_{ij} = -V \langle \hat{c}_{i\downarrow} \hat{c}_{i\uparrow} \rangle \delta_{ij}. \quad (6.9)$$

The effect of the local interactions within the mean-field treatment is thus to add two terms that are diagonal in site space: U, which acts like a regular potential, and Δ, which is some kind of anomalous potential. We shall refer to Δ as the *pairing potential*.

We can rewrite Eq. (6.8) more transparently by writing it in matrix form:

$$\hat{H}_{\text{eff}} = \frac{1}{2} \sum_{ij} \hat{\mathbf{c}}^{\dagger}_i \mathbf{H}_{ij} \hat{\mathbf{c}}_j \quad (6.10)$$

where

$$\hat{\mathbf{c}}_j = \begin{pmatrix} \hat{c}_{j\uparrow} \\ \hat{c}^{\dagger}_{j\downarrow} \\ \hat{c}_{j\downarrow} \\ -\hat{c}^{\dagger}_{j\uparrow} \end{pmatrix} \quad (6.11)$$

is a vector of creation/annihilation operators for electrons with spin up/down on site j, and $\mathbf{H}_{ij}$ is a matrix

$$\mathbf{H}_{ij} = \begin{pmatrix} \tilde{h}_{ij} & \Delta_{ij} & 0 & 0 \\ \Delta^*_{ij} & -\tilde{h}^*_{ij} & 0 & 0 \\ 0 & 0 & \tilde{h}_{ij} & \Delta_{ij} \\ 0 & 0 & \Delta^*_{ij} & -\tilde{h}^*_{ij} \end{pmatrix} \quad (6.12)$$

with $\tilde{h}_{ij} = h_{ij} - E_F \delta_{ij} + U_{ij}$. The minus sign is needed in the definition of $\hat{\mathbf{c}}_j$ because of the anticommutation of the constituent fermionic operators (this can be seen by explicitly expanding out Eq. (6.10)). Writing everything in this form makes explicit the fact that our Hamiltonian is still a quadratic form that can be diagonalised by a unitary transformation. The only difference from the "usual" case is that the unitary transformation will *mix* the $\hat{c}$ and $\hat{c}^{\dagger}$. Also it is made explicit that the Hamiltonian is block diagonal, and that each of these blocks is exactly the Bogoliubov-de Gennes

Hamiltonian of Eq. (6.1).[2] Explicitly, the so-called *Bogoliubov* transformation [6–8],
that diagonalises Eq. (6.8) is

$$
\hat{\beta}_{n\uparrow}^{\dagger} = \sum_j u_{nj} \hat{c}_{j\uparrow}^{\dagger} + v_{nj} \hat{c}_{j\downarrow}
$$
$$
\hat{\beta}_{n\downarrow}^{\dagger} = \sum_j u_{nj} \hat{c}_{j\downarrow}^{\dagger} - v_{nj} \hat{c}_{j\uparrow}
$$
(6.13)

where the $\hat{\beta}^{\dagger}$ are creation operators for *Bogoliubov quasiparticles* that (by definition) are energy eigenstates of the $\hat{H}_{\text{eff}}$, and the u_{nj} and v_{nj} are complex numbers
that satisfy the relation $\sum_j u_{mj}^* u_{nj} + v_{mj}^* v_{nj} = \delta_{nm}$, which ensures that the $\hat{\beta}^{\dagger}$ satisfy fermionic anticommutation relations. By inverting Eq. (6.13) and inserting the
expressions for the $\hat{c}^{\dagger}$ and $\hat{c}$ into Eq. (6.8) we obtain a set of linear equations for
the u_{nj} and v_{nj}. The obtained linear equations are exactly the BdG equation, where
$\mathbf{u}_n = (u_{n0}, \ u_{n1}, \ \ldots)^T$ and $\mathbf{v}_n = (v_{n0}, \ v_{n1}, \ \ldots)^T$. We now also see from Eq.
(6.13) that the u_{nj} are the amplitudes for electron creation, whereas the v_{nj} are the
amplitudes for electron annihilation (i.e. hole creation), which justifies the identification of $\mathbf{u}$ and $\mathbf{v}$ made in Sect. 6.1.

Formally we should determine Δ_{ij} and U_{ij} self-consistently from the obtained
solutions and Eq. (6.9) (see [6] for details). In practice we shall assume that the
effect of U_{ij} is already included into our model for the non-interacting Hamiltonian
elements. In addition, as we shall be dealing with superconductivity in the electrodes
of our system, we shall treat Δ_{ij} as being spatially invariant there, and 0 outside of
the electrodes. Δ_{ij} shall thus be a parameter of our model.

6.2   Relevant Concepts for Superconducting Junctions

In this section we shall introduce the concept of Andreev reflection from a normal-
superconductor boundary. This phenomenon will be the fundamental ingredient necessary to understanding the behaviour of Josephson junctions (a normal region sand-
wiched between two superconductors). We will see how Andreev bound states within
a Josephson junction give rise to a current, even in equilibrium, and how a voltage
bias across a Josephson junction leads to the Hamiltonian of the junction being
necessarily time-dependent.

---

[2]Although in the case where the Hamiltonian is not spin independent the two spin blocks will not
be identical, and in the case where the Hamiltonian is not diagonal in spin space the Hamiltonian
will no longer be block diagonal either.

6.2.1 *Andreev Reflection*

Andreev reflection is a process that occurs at the boundary between a normal material and a superconducting one, where an electron (hole) incident from the normal material is reflected as a hole (electron). By solving a scattering problem (see Sect. 3.2) with a normal electrode and a superconducting one it can be shown [9–11] that the amplitude for an electron of energy E to be reflected from the superconductor as a hole is

$$r_{he}(E) = e^{-i\varphi}\left(\frac{E}{\Delta_0} - i\frac{\sqrt{\Delta_0^2 - E^2}}{\Delta_0}\right) \tag{6.14}$$

where Δ_0 and φ are the magnitude and phase of the pairing potential in the superconductor. For $E < \Delta_0$ (E measured from the Fermi energy) we have $|r_{he}| = 1$, i.e. an incident electron is always reflected as a hole when it has an energy less than the superconducting gap. Figure 6.2 shows an illustration of this process.

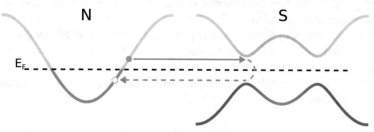

(a) *Semiconductor Picture*: electrons at energy $E > E_F$ are reflected as holes in the Fermi sea with energy $E < E_F$.

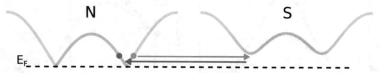

(b) *Excitation Picture*: excitations on the electron branch are reflected onto the hole branch with the same energy.

Fig. 6.2 Two equivalent views of the Andreev reflection process between a normal (N) electrode and a superconducting (S) one

6.2.2 *Josephson Junctions*

Having established Andreev reflection we shall now look at a Josephson junction formed from two superconducting electrodes separated by a piece of normal material. We shall initially consider the case where there is no bias voltage across the junction, but the pairing potential in the left superconductor has a phase φ with respect to the right one. As the left and right superconductors both have a gap of $\pm\Delta_0$ around the Fermi level, any states of the isolated central region with energies inside the gap of the electrodes will be bound states of the fully connected system. We can imagine these bound states as being formed from the successive possible Andreev reflection processes at the left and right electrodes. The spectrum of the bound states can be obtained from a scattering approach [12]. What is interesting about the Josephson junction is that the bound states actually carry a current, which varies with the phase difference φ across the junction. In fact, for short Josephson junctions at equilibrium and zero temperature the bound states are the only part of the spectrum that carries a current; the continuum contribution from the left and right superconductors cancel. This equilibrium current is known as the *d.c. Josephson effect*. Figure 6.3b shows a numerical calculation of the d.c. Josephson effect in a 1D Josephson junction for various phase differences. The current was computed by obtaining the bound state wavefunction of the junction for $E < 0$ (see Sect. 3.1.1, we used 2000 sites in each of the superconducting regions) and applying Eq. (6.6) in the normal part. In this calculation there was a single normal site in the superconducting region, leading to a single pair of bound states in the gap (at energies $+E_b < 0$ and $-E_b > 0$). Figure 6.3b also compares the calculation of the current using Eq. (6.6) with a calculation using the relation

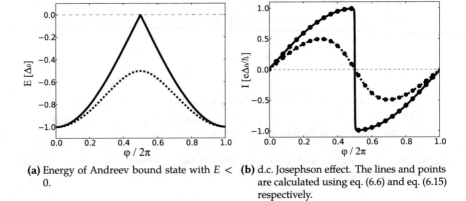

(a) Energy of Andreev bound state with $E < 0$.

(b) d.c. Josephson effect. The lines and points are calculated using eq. (6.6) and eq. (6.15) respectively.

Fig. 6.3 Bound state energy and current for a 1D Josephson junction with a single normal site and phase bias φ at zero temperature. *Solid* and *dashed lines* are for junctions with transmissions of 1 and 0.75 respectively; the *grey dashed lines* are guides for the eye

$$I = \frac{2e}{\hbar} \frac{\partial E_b}{\partial \varphi}. \tag{6.15}$$

This is a special case [13] of the more general relation

$$I = \frac{2e}{\hbar} \frac{\partial F}{\partial \varphi}, \tag{6.16}$$

where F is the free energy, applied to the Josephson junction at zero temperature.

Junctions Under Voltage Bias

Next we shall look at the behaviour of a Josephson junction when a constant bias is applied to one of the superconducting contacts. We shall see that even when the applied bias is time *independent* the Hamiltonian describing the junction is time *dependent*. This in turn leads to a *time-dependent* current flowing across the junction; this is known as the a.c. Josephson effect. We will numerically simulate this in Sect. 6.3.1. In order to treat this situation properly we have to go back to Eq. (6.8). If we write down the mean-field Hamiltonian for a 1D junction with a voltage drop at the interface between the left superconductor and the central (normal) region (we shall take this interface to be between sites 0 and 1) we get[3]

$$\hat{H} = \sum_{ij} \left[\sum_{\sigma} e^{i\varphi(t)\delta_{-1i}\delta_{0j}} \left(h_{ij} - E_F \delta_{ij} + U_{ij} \right) \hat{c}_{i\sigma}^\dagger \hat{c}_{j\sigma} + \right.$$
$$\left. \Delta_0(\theta_{-i} + \theta_{i-l})\delta_{ij}\hat{c}_{i\uparrow}^\dagger \hat{c}_{j\downarrow}^\dagger + \Delta_0^*(\theta_{-i} + \theta_{i-l})\delta_{ij}\hat{c}_{j\downarrow}\hat{c}_{i\uparrow} \right] \tag{6.17}$$

where $\varphi(t) = (e/\hbar) \int V_b dt$ and V_b is the bias voltage, the normal region consists of sites 0 to l inclusive, and θ_m is a discrete Heaviside function, defined as 1 if $m > 0$ and 0 otherwise. Figure 6.4 illustrates this Hamiltonian as a "ladder" consisting of electron and hole sites. If we now apply a gauge transformation[4]

$$\hat{U}(t) = \prod_{j<0} e^{i\varphi(t)\left(\hat{c}_{j\uparrow}^\dagger \hat{c}_{j\uparrow} + \hat{c}_{j\downarrow}^\dagger \hat{c}_{j\downarrow}\right)} \tag{6.18}$$

then our transformed Hamiltonian is

---

[3] Expressed in a gauge where the scalar potential is zero and the vector potential is non-zero exactly at the voltage drop.

[4] This corresponds to a transformation into the Coulomb gauge, where the vector potential will be zero and the scalar potential is non-zero in the left superconductor.

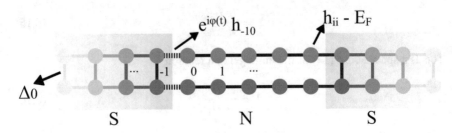

Fig. 6.4 Illustration of the Hamiltonian Eq. (6.17), showing the superconducting (S) and normal (N) regions, as well as the hopping modified by the bias voltage on the *left* (between sites 0 and 1). The *upper* and *lower* chains of sites correspond to electron and hole degrees of freedom respectively. The system extends to infinity on the *left* and *right*

$$\hat{H}' = \sum_{ij}\left[\sum_{\sigma}\left(h_{ij} - E_F\delta_{ij} + U_{ij} + eV_b\delta_{ij}\theta_{-i}\right)\hat{c}_{i\sigma}^\dagger\hat{c}_{j\sigma} + \right.$$
$$\left. \Delta_0(\theta_{-i}e^{-2i\varphi(t)} + \theta_{i-l})\delta_{ij}\hat{c}_{i\uparrow}^\dagger\hat{c}_{j\downarrow}^\dagger + \Delta_0^*(\theta_{-i}e^{2i\varphi(t)} + \theta_{i-l})\delta_{ij}\hat{c}_{j\downarrow}\hat{c}_{i\uparrow}\right]. \tag{6.19}$$

In addition to the usual onsite term eV_b we see that the voltage bias also causes the superconducting terms in the left superconductor to pick up a *time-varying phase*. We can see how the a.c. Josephson effect arises from this by considering the case where the bias voltage is small compared to Δ_0. In this case we can look at the adiabatic regime where the only effect of the bias is to modify the phase across the junction (and we assume that the junction remains in its equilibrium state at all times). At constant bias voltage the phase evolves linearly in time: $\varphi(t) = 2eV_bt/\hbar$. As $I(\varphi)$ oscillates with φ (see Fig. 6.3b), this gives rise to a current that *oscillates in time* with angular frequency $2eV_b/\hbar$: the a.c. Josephson effect.

When we leave the adiabatic regime we can no longer consider the junction to be in a quasi-equilibrium state. One way to picture what happens in the junction in such a regime is to think about the multiple Andreev reflection (MAR) processes that occur when particles traverse the junction. This is easiest to visualise in the semiconductor picture, as shown in Fig. 6.5. We see that at large bias $eV_b > 2\Delta_0$ quasiparticles can tunnel directly from the "valence" band of the left superconductor to the "conduction band" of the right superconductor. As the bias is lowered so that $\Delta_0 < eV_b < 2\Delta_0$ the direct tunneling process is no longer possible, but an electron-like (hole-like) excitation can tunnel into the junction from the left "valence band", be Andreev reflected on the right and tunnel back into the filled "valence band" as a hole-like (electron-like) excitation. As the bias is further lowered into the region $\Delta_0/2 < eV_b < \Delta_0$ the process with a single Andreev reflection no longer becomes possible, and an excitation must undergo two Andreev reflections (ending in the right superconductor) to escape the junction. We can see that whenever eV_b is increased past $2\Delta_0/n$ ($n \in \mathbb{Z}$) that a new process with only n Andreev reflections becomes available. This opening of new "paths" through the junction gives rise to kinks in

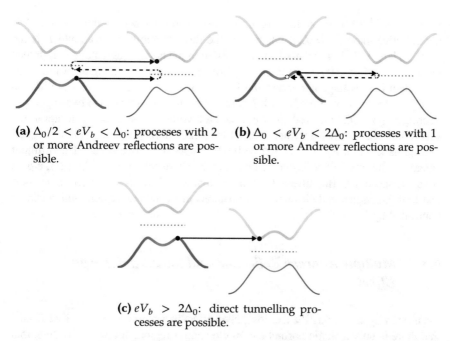

(a) $\Delta_0/2 < eV_b < \Delta_0$: processes with 2 or more Andreev reflections are possible.

(b) $\Delta_0 < eV_b < 2\Delta_0$: processes with 1 or more Andreev reflections are possible.

(c) $eV_b > 2\Delta_0$: direct tunnelling processes are possible.

Fig. 6.5 Visualisation of the multiple Andreev reflection (MAR) processes in a biased Josephson junction. As the bias is raised processes with fewer and fewer Andreev reflections become available, until the bias voltage eV_b exceeds $2\Delta_0$, twice the superconducting gap, when quasiparticles in the "valence band" of the *left* superconductor can directly tunnel into the "conduction band" of the *right* superconductor

the sub-gap current-voltage characteristic of a Josephson junction, as we shall see in Sect. 6.3.1. This picture will also be used in Sect. 6.3.2 to explain the relaxation of Andreev bound states in the non-equilibrium regime.

6.3 Time-Resolved Simulations of Josephson Junctions

We are now in a position where we can perform time-resolved simulations of Josephson junctions and be able to interpret our results within the framework of the concepts introduced above. We shall use the same basic model of a 1D SNS junction for all the simulations presented in this section. We can write down the elements of the BdG Hamiltonian compactly as:

$$\begin{aligned}
H_{j,j} &= \left[2\gamma - E_F + U(\theta_{l-j} - \theta_{-j})\right] \\
\Delta_{j,j} &= \Delta_0(\theta_{-j} + \theta_{j-l}) \\
H_{j,j+1}(t) &= -\gamma \exp\left[i\varphi(t)\delta_{-1,j}\right]
\end{aligned} \tag{6.20}$$

where $\gamma = \hbar^2/(2m^*a^2)$ is the hopping parameter of the discretised problem (m^* is the effective mass), $l = L/a$ where a is the discretisation step used and L is the junction length. U is a potential barrier present in the normal region that will induce normal scattering. $\varphi(t) = (e/\hbar) \int_0^t V_b(t')dt'$ is, as usual, the integral of the time-dependent bias voltage applied to the left electrode. We see that this Hamiltonian describes a normal region of length L (our central region) attached to two superconducting regions (the electrodes). We model the voltage drop as being abrupt at the superconductor-normal interface. For all the simulations presented below we shall use the parameters $E_F = \gamma$ and $\Delta = 0.02\gamma$ unless otherwise stated. The different situations that we will consider will correspond to: different forms for $\varphi(t)$, e.g., a constant bias or a pulse; different lengths L, that will allow us to consider both short and long junctions; and different transmissions of the central region, which can be controlled by U.

6.3.1 Multiple Andreev Reflection and the a.c Josephson Effect

Let us start by simulating the a.c. Josephson effect in a short junction. We set L such that there is only a single normal site in the central region, and we set U such that the transmission through the junction is 0.7. We use a bias voltage of the form

$$V_b(t) = \begin{cases} 0, & t < 0 \\ \frac{V_b}{2}\left(1 - \cos(\frac{\pi t}{T})\right), & 0 \le t \le T \\ V_b, & t > T. \end{cases} \qquad (6.21)$$

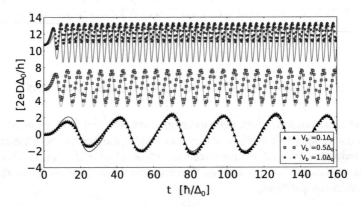

Fig. 6.6 The a.c. Josephson effect. The different *curves* show the calculated current as a function of time for different bias voltages across a short junction with a transmission of 0.7. The *solid curves* and *symbols* show the theoretical and numerical results respectively. The *curves* have been vertically offset for clarity

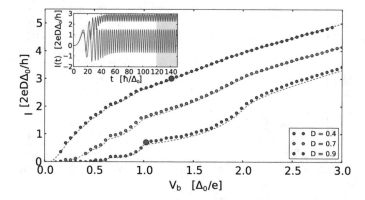

Fig. 6.7   d.c. current-voltage curve showing the analytical results from ref. [14] (*dashed line*) and the source-sink numerical calculation (*points*) for different values of the transmission (*D*) of the insulating link. Inset: time series corresponding to the *enlarged points* in the main figure, showing a typical averaging window over which the d.c. current was calculated

We see that this corresponds to raising the bias voltage from 0 to V_b during a time T ($\sim \hbar/\Delta_0$). The reason that this procedure is necessary is because of the presence of Andreev bound states (see Sect. 3.1.1).

Figure 6.6 shows the current calculated at the interface between the left contact and the central region, compared to the adiabatic result discussed in Sect. 6.2.2 (the spectrum $E(\varphi)$ was calculated from the equilibrium junction and differentiated numerically). We see that the two results agree at low bias, but that the adiabatic result is not sufficient to capture the behaviour in the strong bias regime; hence the added value of numerical techniques in such a regime.

As we increase the bias we see a d.c. component appearing in the current. This is due to the MAR processes discussed in Sect. 6.2.2. With our simulations we can go beyond the low bias regime and look at the full current-voltage characteristic, which should allow us to see the effect of the MAR even more clearly. The Fourier components of the MAR current have been previously calculated using a Floquet approach [14, 15], and are routinely observed experimentally (see for instance ref. [16]). More recently, some numerical results were obtained with techniques working in the time domain [1, 3]; our simulations follow a similar line here. We perform a series of simulations for different bias voltages and calculate the d.c. component of the current by averaging over a number of periods. The results are shown in Fig. 6.7 for different values of the junction transmission (*D*) and are compared with the analytical results of ref. [14]. We see a very good quantitative agreement with these previous results, and nearly perfectly reproduce the "kinks" whenever new MAR processes become possible.

We can also go beyond the limits of analytical approaches for a modest extra computational cost. We can, for example, explore the behaviour of a long Josephson junction under voltage bias. Figure 6.8 compares the current-voltage characteristics of a long junction with the short junction studied previously. We clearly see that

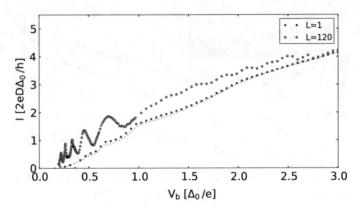

Fig. 6.8 Comparison of the current-voltage characteristics for a *short junction* (one site in the normal region) and a *long junction* (120 sites in the normal region). Both the junctions have a transmission of 0.7

the long junction has more sub-gap features, which can be attributed to the larger number of Andreev states below the gap. We see that numerics has an advantage over analytical approaches in this regard, in that it is relatively cheap to explore new regions of parameter space or in crossover regions between tractable limits (e.g. short junction versus long junction).

6.3.2   Relaxation of Andreev Bound States

The above calculations were performed using the procedure for including the bound state contribution discussed in Sect. 3.1.1. While this contribution is necessary to get the correct transient behaviour of the current, in this section we shall see that in the presence of finite bias the contribution of the initially filled Andreev bound states to the current tends to zero with time. We shall see that this relaxation can be seen to come from MAR processes that are not available in the equilibrium junction. This means that in the presence of finite bias the long-time behaviour of the system will be independent of the initial filling of the Andreev bound state(s).

Suppose that we start in equilibrium at $t = 0$, and at $t = 0^+$ we abruptly raise the bias voltage, thereby placing the system in a non equilibrium state. Just after the voltage rise, a given wave function can be decomposed on the eigenbasis of the equilibrium SNS junction,

$$\psi(0^+) = \sum_\alpha \int dE \, c_{\alpha E} \, \psi_{\alpha E}^{\text{st}} + \sum_n c_n \psi_n^{\text{bnd}}, \tag{6.22}$$

where $c_{\alpha E}$ and c_n are respectively the projection of the wave function on the scattering states ($\psi_{\alpha E}^{\text{st}}$) and the Andreev bound states (ψ_n^{bnd}). It is important to realize that in the

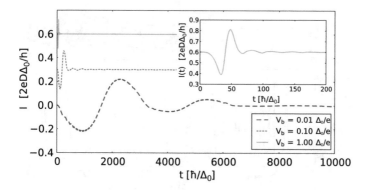

Fig. 6.9 Current contribution from the (Andreev) bound states at different bias voltages. The curves have been offset for clarity. The inset shows a zoom of the curve for $V_b = \Delta/e$. $\Delta_0 = 0.1\gamma$ for these simulations

absence of bias voltage, the bound state part of the wave function will *never* relax (within the above model) as the Andreev states are true bound states with energy E_n: the second part of the wave function will simply oscillate as $\sum_n c_n e^{-iE_n t} \psi_n^{bnd}$ forever. However, the presence of the bias voltage allows the energy to change by eV in between successive Andreev reflections so that after $N \sim \Delta_0/(eV_b)$ reflections, one can reach energies outside the gap and the wave function can relax. Denoting $\tau_P = L/v_F$ the time of flight between two Andreev reflections, we expect the relaxation time τ_R of the system to behave as $\tau_R \propto N\tau_F = L\Delta_0/(v_F e V_b)$.

Figure 6.9 shows the contribution to the current coming from the initially filled Andreev states as a function of time for different bias voltages. We indeed see that the current carried by the bound states dies away with time in presence of a finite bias. Although we did not define τ_R precisely, we clearly see that dividing V_b by a factor of 10 leads to a tenfold increase in the relaxation time, establishing the relation $\tau_R \propto 1/V_b$ that originates from the MAR assisted relaxation process.

6.3.3 Propagation of a Voltage Pulse

A natural consequence of the above discussion is that if one sends a fast voltage *pulse* through the system (i.e. the final bias voltage vanishes instead of having a finite value), then the corresponding bound state contribution will not relax and will oscillate forever (within the assumptions of our model).

Long Junction

We shall first look at a transparent ($U = 0$), long junction and apply a Gaussian voltage pulse of duration (full-width at half-max) τ_p on the left superconductor. The time of flight through the junction at the Fermi energy is then $\tau_F = L/v_F$ where v_F is the Fermi velocity. Our junction is "long" in the sense that $\Delta_0\tau_F/\hbar \gg 1$ (it consists

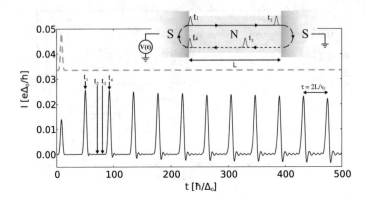

Fig. 6.10 Current (*solid line*) and voltage (*dashed line*, offset for clarity) at the left superconducting-normal contact as a function of time. Inset: propagation of the charge pulse through the junction at different times (t_1, t_2, t_3, t_4) and the corresponding times indicated on the main plot

of 350 sites in the central region, and we used $\Delta_0 = 0.1$ for these simulations), and we will look at "fast" pulses where $\tau_F/\tau_p \gg 1$ ($\tau_F/\tau_p \sim 5$ in our case). Intuitively we can have the following picture of how the system evolves after the voltage pulse is applied. The pulse generates an electron-like excitation that propagates through the system until it reaches the right superconductor. There, it is Andreev reflected as a hole-like excitation. The excitation now propagates backward towards the left superconducting electrode where it is Andreev reflected a second time. As the voltage pulse was fast with respect to the time of flight, the voltage on the left electrode is again 0 when the excitation reaches it. This means that upon Andreev reflection the excitation does not pick up any energy (as it would if a finite bias were applied). Consequently the excitation cannot escape the junction and continues to oscillate back and forth forever. This is rather appealing: one sends a short voltage pulse and gets an oscillating current at frequency $1/(2\tau_F)$. Beyond the current model, the relaxation time of the system will be given by the fluctuations of the voltage due to the electromagnetic environment and we anticipate a relaxation of the current on a scale given by the corresponding RC time.

Figure 6.10 shows a numerical simulation of the propagation of a voltage pulse as discussed above. Despite the fact that there is only a single voltage pulse at the start, we see pulses of current every $2\tau_F$. We do not observe any quasiparticle current in the superconducting lead; this (super)current is purely associated with the Andreev reflection process described above.

We can go a little bit further and look at the structure of the bound states that carry the supercurrent. They are given by the stationary condition [12]

$$r_A(E)^2 e^{2iE\tau_F/\hbar} e^{i\varphi} = 1, \tag{6.23}$$

where the left-hand superconductor is at a phase bias φ compared to the right-hand one and $r_A(E)$ is the Andreev reflection amplitude given in Sect. 6.2. The paths

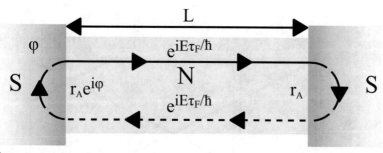

(a) Paths with right-going electrons.

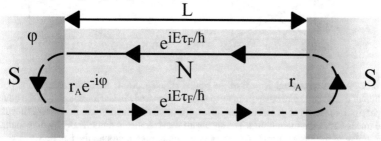

(b) Paths with left-going electrons.

Fig. 6.11 Sketches of the two classes of paths that can result in bound states. The *solid lines* corresponds to an electron-like excitation, and the *dashed lines* to a hole-like one. Andreev reflection at the normal-superconductor interface converts an electron-like excitation to a hole-like one. Each sketch actually represents a set of paths with 1, 2, 3, ... pairs of Andreev reflections

contributing to this amplitude are sketched in Fig. 6.11a. A similar expression exists for the reversed paths where the sign of φ is flipped; this is sketched in Fig. 6.11b. For $E < \Delta_0$ we can re-write this condition as

$$- 2 \arccos(E/\Delta_0) + \frac{2E\tau_F}{\hbar} \pm \varphi = 2\pi m , \quad m \in \mathbb{Z}. \tag{6.24}$$

In the long junction limit ($\Delta_0 \gg \hbar/\tau_F$) close to zero energy this simplifies to:

$$E = \frac{h}{2\tau_F} \left[m + \frac{1}{2} \mp \frac{\varphi}{2\pi} \right] \tag{6.25}$$

which corresponds to two sets of equidistant energies separated by $h/(2\tau_F)$: one set has energy increasing with φ, and the other decreasing with φ. Each of these sets corresponds to ballistic propagation in the continuum limit $\tau_P \ll \tau_F$. The numerical spectrum, which is shown in Fig. 6.12, adheres to the above-derived result except near the degeneracy points. The degeneracies are lifted due to the finite ratio Δ_0/E_F used in the numerical calculation, which induces a finite normal reflection at the normal-superconducting interfaces. The two insets of Fig. 6.12 show two time dependent

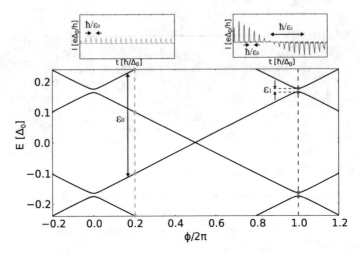

Fig. 6.12 A section around $E = 0$ of the bound state spectrum after the passage of a pulse as a function of the phase ϕ picked up from the pulse. The *vertical dashed lines* highlight the bound state energies for two values of ϕ. The current flowing through the junction as a function of time is shown in the traces above the main figure. The spectrum was calculated numerically by diagonalizing the Hamiltonian of the system projected onto a large, finite region around the junction

simulation at two different values of the superconducting phase difference *after* the pulse, $\varphi = \varphi(t = \infty)$. We see that when the two sets of bound states are very close in energy the output current beats with a frequency which is given by the level spacing. For well-spaced bound states this frequency is so high that it has no visible effect on the current trace.

Short Junction

The above effect is intriguing, but unfortunately long ballistic Josephson junctions are difficult to realize experimentally (with the exception perhaps of carbon nanotubes). In diffusive junctions there will be a distribution of times of flight which will wash out the above effect. An alternative is to consider the limit of short junctions, which have been studied extensively experimentally with atomic contacts (break junctions) [16]. We shall, therefore, now explore the effect of a voltage pulse applied to a short Josephson junction. We do not expect to be able to see a train of well-resolved peaks of current, as in the long junction case, because the time of flight of the short junction is much shorter than the typical pulse duration. We do, however, expect to see the effect that gives rise to the "beating" in Fig. 6.12, as this is governed only by the energy difference between the Andreev bound states in the junction. Figure 6.13 shows the current passing through a short junction when voltage pulses of varying heights are applied. We see an initial transient part followed by an oscillatory part that continues indefinitely. Initially, all the states up to $E = 0$ are filled. The pulse excites some quasiparticles into states at $E > 0$ and also shifts the phase bias across the junction so that we are at a different place in the phase-energy plot than we were before the pulse

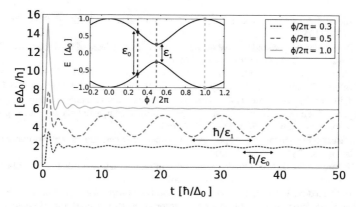

Fig. 6.13 Current traces as a function of time for three different voltage pulses applied to a short Josephson junction with a transparency of 0.9. The *curves* have been offset for clarity. Each pulse has a full-width half maximum of $0.4\,\hbar/\Delta_0$, and the pulses are of different heights. This gives a different phase bias, ϕ, across the junction after the pulse has completed. Inset: The bound state spectrum for the junction as a function of the phase bias, the phases accumulated by the three pulses are indicated by *dashed lines*

(indicated by dashed lines in the inset to Fig. 6.13). Any quasiparticles in continuum states escape into the leads after some time ($\sim 20\hbar/\Delta_0$ in Fig. 6.13), however the contribution in the Andreev bound states cannot escape. After we have reached a steady state we are essentially in a superposition of Andreev bound states at energy E and $-E$. These two contributions interfere with one another to give a current that *oscillates* in time at angular frequency $2E/\hbar$. This effect is most strongly seen for $\varphi = \pi$, as the Andreev levels have the smallest energy gap here. For $\varphi = 2\pi$ the oscillations die away with time, as the Andreev levels hybridize with the continuum at this point. By tuning the energy gap between the Andreev levels after the pulse we are able to control the frequency of the current. We can tune the energy gap by placing ourselves at different points in the phase-energy diagram (by sending in pulses of different heights), or by tuning the transparency of the junction to modify the phase-energy diagram itself.

References

1. E. Perfetto, G. Stefanucci, M. Cini, Equilibrium and time-dependent Josephson current in one-dimensional superconducting junctions. Phys. Rev. B **80**(20) (2009)
2. J. Weston, X. Waintal, A linear-scaling source-sink algorithm for simulating time resolved quantum transport and superconductivity, arXiv:1510.05967 [cond-mat] (Oct. 2015)
3. G. Stefanucci, E. Perfetto, M. Cini, Time-dependent quantum transport with superconducting leads: a discrete-basis Kohn-Sham formulation and propagation scheme. Phys. Rev. B **81**(11), 115446 (2010)
4. J. Bardeen, L.N. Cooper, J.R. Schrieffer, Microscopic theory of superconductivity. Phys. Rev. **106**(1), 162–164 (1957)

5. J. Bardeen, L.N. Cooper, J.R. Schrieffer, Theory of superconductivity. Phys. Rev. **108**(5), 1175–1204 (1957)
6. P.G. De Gennes, Superconductivity of metals and alloys. Advanced book classics. (Reading, Mass: Advanced Book Program, Perseus Books, 1999)
7. N.N. Bogoliubov, A new method in the theory of superconductivity I. JETP 7.1, **41** (1958)
8. J.G. Valatin, Comments on the theory of superconductivity. Nuovo Cim **7**(6), 843–857 (1958)
9. G.E. Blonder, M. Tinkham, T.M. Klapwijk, Transition from metallic to tunneling regimes in superconducting microconstrictions: excess current, charge imbalance, and supercurrent conversion. Phys. Rev. B **25**(7), 4515–4532 (1982)
10. C.W.J. Beenakker, Quantum transport in semiconductor-superconductor microjunctions. Phys. Rev. B **46**(19), 12841–12844 (1992)
11. Y.V. Nazarov, Y.M. Blanter, Quantum transport: introduction to nanoscience. (Cambridge University Press, Cambridge, UK, New York, 2009)
12. C.W.J. Beenakker, Three "Universal" mesoscopic Josephson effects, in *Transport Phenomena in Mesoscopic Systems*, vol. 109, ed. by M. Cardona et al. (Springer, Berlin, Heidelberg, 1992), pp. 235–253
13. C.W.J. Beenakker, H. van Houten, The superconducting quantum point contact, in *Nanostructures and Mesoscopic Systems* (Academic Press, 1992), pp. 481–497
14. D. Averin, A. Bardas, ac Josephson effect in a single quantum channel. Phys. Rev. Lett. **75**(9), 1831–1834 (1995)
15. J.C. Cuevas, A. Martín-Rodero, A. Levy Yeyati, Hamiltonian approach to the transport properties of superconducting quantum point contacts. Phys. Rev. B **54**(10), 7366-7379 (1996)
16. E. Scheer et al., Conduction channel transmissions of atomic-size aluminum contacts. Phys. Rev. Lett. **78**(18), 3535–3538 (1997)

Chapter 7
Manipulating Andreev and Majorana Resonances in Nanowires

In Chap. 6 we studied the effect of a voltage pulse on a Josephson junction, and saw several interesting effects due principally to the unique role that the Andreev bound states play in superconducting systems. Even earlier in Chap. 5 we saw how the electronic wavefunction had its phase *twisted* by the application of a voltage pulse, and that this leads to a dynamical modification of any interference in the system. In this chapter we shall connect these two ideas to study the effect of a *train* of voltage pulses applied to a normal-insulator-normal-superconducting (NINS) junction; we shall see how this leads to a *steady state* modification of the interference. We shall then turn to a more involved model that exhibits exotic Majorana excitations, and will show that we can use the same trains of voltage pulses to manipulate them. Finally we shall show that we can identify Andreev and Majorana states by "spectroscopy" in the presence of a train of voltage pulses. This may have implications for detection of Majorana states in recent experimental setups [1].

7.1 Model for Normal-Insulator-Normal-Superconductor (NINS) Junctions

In this section we shall present the general system under study, and the model that we shall use in the numerical simulations that follow. Figure 7.1 shows a sketch of the system: a one-dimensional wire consisting of a normal region (N) coupled to a superconductor (S), with an insulating barrier (I) in the normal region at a length L_J from the normal-superconductor interface. The height of the insulating barrier can be controlled with a gate voltage V_T, and the whole junction has a bias $V_b + V_P(t)$ applied. The Hamiltonian for this model is:

$$\hat{H}(t) = \int_{-\infty}^{\infty} dx \, \hat{\boldsymbol{\Psi}}^{\dagger}(x) \mathbf{H}_{\mathrm{BdG}}(x, t) \hat{\boldsymbol{\Psi}}(x) \tag{7.1}$$

© Springer International Publishing AG 2017
J. Weston, *Numerical Methods for Time-Resolved Quantum Nanoelectronics*,
Springer Theses, DOI 10.1007/978-3-319-63691-7_7

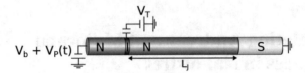

Fig. 7.1 Sketch of the junction to be studied. A section of normal material (N) coupled to a superconducting region (S), with an insulating barrier (I) at a distance L_J from the NS junction. The height of the insulating barrier can be controlled with a gate voltage V_T, and the junction has a time-independent bias V_b applied, as well as a time-dependent component $V_P(t)$

where x is the position along the nanowire,

$$\hat{\Psi}(x) = \begin{pmatrix} \hat{\psi}_\uparrow(x) \\ \hat{\psi}_\downarrow(x) \\ \hat{\psi}_\uparrow^\dagger(x) \\ -\hat{\psi}_\downarrow^\dagger(x) \end{pmatrix} \tag{7.2}$$

is a vector of creation/annihilation operators that create/destroy particles with spin up/down at position x and $\mathbf{H}_{\text{BdG}}(x, t)$ is the Bogoliubov-de Gennes Hamiltonian given by

$$\mathbf{H}_{\text{BdG}}(x, t) = \left[\frac{-\hbar^2}{2m^*} \frac{\partial^2}{\partial x^2} + eV_T\delta(x) + eV_P(t)\Theta(-x) - E_F \right] \tau_3 + \Delta_0\Theta(x - L_J)\tau_1 \tag{7.3}$$

where m^* is the effective mass, e is the electronic charge, Δ_0 is the superconducting order parameter, E_F is the Fermi energy, and $\theta(x)$ and $\delta(x)$ are the Heaviside and Dirac delta functions respectively. The τ_n are Pauli matrices that act in particle-hole space; for the above-chosen basis they can be written as

$$\tau_1 = \begin{pmatrix} 0 & 1 \\ 1 & 0 \end{pmatrix} \otimes \mathbb{1}_2, \quad \tau_2 = \begin{pmatrix} 0 & -i \\ i & 0 \end{pmatrix} \otimes \mathbb{1}_2, \quad \tau_3 = \begin{pmatrix} 1 & 0 \\ 0 & -1 \end{pmatrix} \otimes \mathbb{1}_2 \tag{7.4}$$

where $\mathbb{1}_2$ denotes a 2×2 identity matrix and $\otimes$ is the Kronecker product. Initially the problem may seem over-specified—as our Hamiltonian is currently spin independent—however we have started with this sufficiently general formulation so as to introduce the necessary notation for Sect. 7.4.2. As usual we perform a gauge transformation (see appendix C) and discretise this model on a lattice of spacing a to obtain the tight binding model

$$\hat{H}_{\text{tb}} = \sum_j \hat{c}_j^\dagger \mathbf{H}_{j,j} \hat{c}_j + \hat{c}_j^\dagger \mathbf{H}_{j,j+1} \hat{c}_{j+1} + h.c. , \tag{7.5}$$

where $\hat{c}_j = \hat{\Psi}(ja)$ and

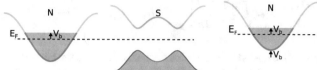

(a) Chemical potential gradient: more states on the left are filled compared with equilibrium.

(b) Electrical potential gradient: states on the left are filled the same as in equilibrium, but the band is shifted.

Fig. 7.2 Comparison of application of electrical or chemical potential across a normal-superconducting junction

$$\mathbf{H}_{j,j} = [2\gamma - E_F + eV_T\delta_{0,j}]\tau_3 + \Delta_0\theta_{j,l_J}\tau_1$$
$$\mathbf{H}_{j,j+1} = -\gamma e^{i\varphi_P(t)\delta_{-1,j}\tau_3}\tau_3 \tag{7.6}$$

where $\gamma = \hbar^2/2m^*a^2$ is the hopping parameter, $l_J = L_J/a$, $\varphi_P(t) = (e/\hbar)\int_0^t V_P(u)\,du$, $\delta_{m,n}$ is the Kronecker delta, and $\theta_{m,n}$ is a discrete Heaviside function, defined as 1 if $m > n$ and 0 otherwise. The astute reader will notice that the static bias V_b does not appear in any of these expressions for the Hamiltonian. The reason for this is that we shall apply the static bias as a modification to the *chemical* potential of the left lead ($x < 0$); this means that it will only enter into the statistical physics part of the problem, when we integrate over all energies when calculating observables. This means that we start in an *out-of-equilibrium* situation at $t = 0$ (finite bias), which negates the need to manually raise the bias from zero and wait for the system to relax; this is clearly advantageous from a numerical point of view. Figure 7.2 illustrates the difference between the addition of a purely chemical potential, as opposed to an electrical potential.[1] Concretely this allows us to write the following expression for the time-dependent current, applying Eq. (3.7) to the present case:

$$I(t) = \int \frac{dE}{2\pi} [f(E - eV_b) I_l(E,t) + f(E) I_r(E,t)] \tag{7.7}$$

where $f(E)$ is the Fermi function centred around E_F, and $I_l(E,t)$ and $I_r(E,t)$ are defined by the contributions to the current flowing across the system/lead interface between sites 0 and 1 at injection energy E from scattering states originating in the left (l) and right (r) leads respectively:

$$I_\alpha(E,t) = 2e\Im\{[\psi_{\alpha E}^\dagger(t)]_1\tau_3\mathbf{H}_{1,0}[\psi_{\alpha E}(t)]_0\}, \tag{7.8}$$

where $\mathbf{H}_{1,0}$ is defined by Eq. (7.6), τ_3 is defined by Eq. (7.4), and $[\psi_{\alpha,E}(t)]_j$ is a vector of components of the scattering wavefunction on site j at time t originating in lead α at energy E, in the basis defined by Eq. (7.2). We see that this allows us to write the following expression for the *time-resolved* differential conductance:

---

[1]This difference was extensively studied in Ref. [2].

$$\frac{\partial I(t)}{\partial V_b} = - \int \frac{dE}{2\pi} \left. \frac{df}{dE} \right|_{E - eV_b} I_l(E, t),$$ (7.9)

where primes denote derivatives. As df/dE is sharply peaked around the Fermi level (in the limit of zero temperature $df/dE \rightarrow \delta(E - E_F)$) this means that calculating the time-resolved differential conductance is computationally much cheaper than calculating the full current, as only a relatively small energy window needs to be integrated (and only a single energy $E_F + V_b$ is needed in the zero-temperature limit). In all the cases that follow we are going to consider a *periodic* driving $V_P(t)$ (a train of voltage pulses, for example), so the output current $I(t)$ will have the same periodicity in the steady state. Assuming that $I(t)$ has period T after some time t', we can define a more experimentally relevant quantity, the time average of the time-resolved differential conductance:

$$\left\langle \frac{\partial I(t)}{\partial V_b} \right\rangle_T = -\frac{1}{T} \int \frac{dE}{2\pi} \left. \frac{df}{dE} \right|_{E - eV_b} \int_{t'}^{t'+T} dt\, I_l(E, t).$$ (7.10)

This corresponds to a d.c. measurement of the differential conductance in the presence of the periodic driving $V_P(t)$. The time average is taken at a time t' sufficiently large that the system will have reached a steady state: Even though we start at finite bias at $t = 0$, our approach still requires that $V_P(t)$, the time-dependent part of the bias, be 0 at $t \leq 0$ so the system will take a finite time to reach a steady state.

7.2   Scattering Approach to Andreev Resonances

In Sect. 6.2 we introduced the notion of Andreev reflection, when an electron is reflected as a hole from a normal-superconducting (NS) interface. In our NINS junction the presence of the insulating barrier provides a further source of reflection (normal reflection this time, as opposed to Andreev reflection). This already gives us a hint that there are going to be resonances present in the system due to the coherent superposition (and interference) of the different paths through the system. In this way scattering theory provides an intuitive method of for understanding these so-called "Andreev" resonances in terms of an effective Fabry-Perot interferometer.

First we write down the generalised Landauer formula to calculate the d.c. conductance G for our NINS junction [3–5]:

$$G = \frac{e^2}{h} \sum_{\sigma\sigma'} \left[\delta_{\sigma\sigma'} + \left| r_{h\sigma', e\sigma} \right|^2 - \left| r_{e\sigma', e\sigma} \right|^2 \right]$$ (7.11)

where $r_{h\sigma', e\sigma}$ ($r_{e\sigma', e\sigma}$) is the amplitude for an electron with spin σ ($\sigma \in \{\uparrow, \downarrow\}$) incident from the normal lead on the left to be reflected as a hole (electron) with spin σ'. In the present case, where the Hamiltonian is spin independent, the only spin-flip process is during the Andreev reflection, so $r_{h\uparrow,e\uparrow} = r_{e\downarrow,e\uparrow} = 0$ (similarly when $\uparrow$ and $\downarrow$ are switched). In addition, if we consider energies below the gap $E < \Delta_0$ and use the fact that the number of quasiparticles is conserved, $\left|r_{h\downarrow,e\uparrow}\right|^2 + \left|r_{e\uparrow,e\uparrow}\right|^2 = 1$ (similarly when $\uparrow$ and $\downarrow$ are switched), we can simplify this to

$$G = \frac{4e^2}{h}\left|r_{h\downarrow,e\uparrow}\right|^2. \tag{7.12}$$

We shall now calculate $r_{h\downarrow,e\uparrow}$ by using an approach equivalent to that used in Refs. [3, 6]. Instead of reasoning directly in terms of scattering matrices, however, we will instead explicitly sum over the amplitudes for different possible paths through the system, similar to the approach used in Ref. [7]. This more intuitive approach will also simplify the treatment when we go to the time-dependent case. First we shall define the amplitudes for the constituent processes. We shall denote $s(E) = e^{ik(E)L_J}$ the amplitude for free propagation of an electron with energy E (relative to the Fermi energy) from one side of the junction to the other.[2] As we will be concerned with energies smaller than (or of the order of) the superconducting gap Δ_0, and as $\Delta_0 \ll E_F$, we can approximate the spectrum as being linear around the Fermi wavevector k_F: $k(E) \approx k_F + E/\hbar v_F$, where v_F is the Fermi velocity. This allows us to write $s(E) = e^{iE\tau_F/\hbar}$, where $\tau_F = L_J/v_F$ is the time of flight through the junction, and we have removed the $e^{ik_F L_J}$ factor, which will just give a global phase. The equivalent amplitude for holes is $[s(-E)]^* = e^{iE\tau_F}$. The Andreev reflection amplitude for a spin up electron to be reflected as a spin down hole is $\rho_{h\downarrow,e\uparrow}(E) = r_A(E) \equiv E/\Delta_0 - i\sqrt{1 - (E/\Delta_0)^2}$, and similarly for the other process in the same spin sector: $\rho_{e\uparrow,h\downarrow}(E) = r_A(E)$. The processes in the other spin sector have an extra minus sign due to the chosen basis (defined by Eq. (7.2)): $\rho_{e\downarrow,h\uparrow}(E) = \rho_{h\uparrow,e\downarrow}(E) = -r_A(E)$. Finally we shall denote $d(E)$ $(r(E))$ the transmission (reflection) amplitude for electrons through the insulating barrier[3]; the equivalent amplitude for holes is $d^*(-E)$ $(r^*(-E))$.

The first three possible paths that take a spin up electron to a spin down hole are sketched in Fig. 7.3, along with their associated amplitudes. The total amplitude can be obtained by summing over all the paths:

$$r_{h\downarrow,e\uparrow} = d^2 r_A \sum_{m=0}^{\infty} r^{2m} r_A^{2m} s^{4m}, \tag{7.13}$$

where we have suppressed the explicit energy dependence for the sake of clarity. This can be resummed as

$$r_{h\downarrow,e\uparrow} = \frac{d^2 r_A}{1 - re^{i\Phi(E)}} \tag{7.14}$$

---

[2]We have dropped the spin index as the transport is spin independent.
[3]We have dropped the spin index as the transport is spin independent.

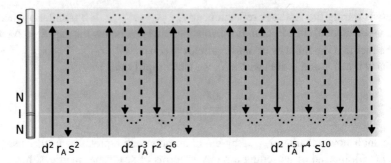

$$d^2\, r_A\, s^2 \qquad\qquad d^2\, r_A^3\, r^2\, s^6 \qquad\qquad d^2\, r_A^5\, r^4\, s^{10}$$

Fig. 7.3 Sketch of the first three paths through the NINS junction that contribute to the amplitude $r_{h\downarrow,\,e\uparrow}$—and their individual amplitudes—which correspond to the first three terms in Eq. (7.13). *Solid lines* correspond to electrons, dashed lines to holes. The explicit energy dependence has been dropped for compactness

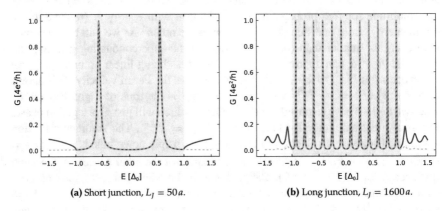

(a) Short junction, $L_J = 50\,a$.   **(b)** Long junction, $L_J = 1600\,a$.

Fig. 7.4 d.c. differential conductance in short and long NINS junctions, calculated numerically using the model Eq. (7.6) with $\gamma = 1$, $E_F = 1$, $\Delta_0 = 0.01$, and $V_T = 3.0$. The *solid curves* were calculated using Eq. (7.11), while the *dashed curves* were calculated using Eq. (7.12). We see that they match for $|E| < \Delta_0$ (*shaded region*), but differ for $|E| > \Delta_0$ due to the quasiparticle current in the superconductor

where

$$\Phi(E) = -2\arccos\left(\frac{E}{\Delta_0}\right) + \frac{4E\tau_F}{\hbar}, \tag{7.15}$$

where we have used $r_A(E) = \arccos(E/\Delta_0)$, which is valid for $E < \Delta_0$. The expression for the other spin sector is almost identical: $r_{h\uparrow,\,e\downarrow} = -r_{h\downarrow,\,e\uparrow}$. Thus we see the analogy between the present system and a Fabry-Perot interferometer. The Andreev resonances[4] (corresponding to maxima of $r_{h\downarrow,\,e\uparrow}(E)$) occur at energies E_q that minimise the denominator of Eq. (7.14), i.e. they satisfy $\Phi(E_q) = 2\pi q$ ($q \in \mathbb{Z}^*$). Figure 7.4 shows this resonant structure in the sub-gap differential conductance.

---

[4]So called because there is an Andreev reflection involved.

7.3 Trains of Voltage Pulses Applied to NINS Junctions

In Sect. 5.3 we saw that applying a voltage pulse puts a *twist* into the phase of the stationary wavefunctions; due to the different propagation times along the different paths the interference pattern is modified during a finite time window, and returns to its d.c. state once the phase twist has arrived from all the paths (only two paths in Chap. 5). Here, we seek to stabilise this transient regime by applying a *train* of voltage pulses. Let us consider an example where we apply a sequence of identical pulses that each induce a phase shift of φ, separated by $4\tau_F$. We have not chosen this time delay arbitrarily; it corresponds to the time difference between successive paths through the system (see Fig. 7.3). This means that when the phase twist from the nth pulse in the sequence is just arriving at the output from the first path, the $(n-1)^{th}$ phase twist is just arriving from the second path, and so on. The result is that the nth path will have a phase φ with respect to the $(n-1)^{th}$ path (compared to the d.c. case) *at all times*, not just during a finite time window. In a similar way we can see that by varying the delay between subsequent voltage pulses, as well as the phase shift they induce (by changing their height or duration), we will be able to modify the interference pattern in different ways.

Let us now be more concrete in our reasoning; we shall consider the case where we apply $V_P(t)$, a T-periodic train of voltage pulses on top of the bias V_b. Following the same reasoning as Sect. 5.3, and considering that the voltage pulses only serve to twist the phase of the stationary wavefunctions, we can write the time-dependent electron-hole reflection amplitude as

$$r_{h\downarrow,e\uparrow}(t, E) = d^2 r_A \sum_{m=0}^{\infty} r^{2m} e^{i\Phi_m(t,E)} \tag{7.16}$$

with

$$\Phi_m(t, E) = m\Phi(E) + \varphi(t + 4\tau_F m) \tag{7.17}$$

where $\varphi(t) = (e/\hbar) \int_0^t V_P(u)\mathrm{d}u$ is the phase picked up due to $V_P(t)$. The arrival of the phase change for the mth order contribution is delayed by $4\tau_F m$ because of the extra distance to traverse in the mth order path. We will consider trains of pulses of different shapes as shown in Fig. 7.6a: a train of upright pulses; a train of alternating pulses; and also a sinusoid, which can be seen as a deformation of the train of alternating pulses. In the following we will denote the phase accumulated over one period $\varphi(T)$ (upright pulses) or half a period $\varphi(T/2)$ (alternating and sine pulses) as $2\pi\bar{n}$.

The simplest situation is when one sends a series of upright localized pulses (of widths much shorter than τ_F). Figure 7.5 shows a snapshot of the different paths at an instant t in time, where such a train of pulses with period $4\tau_F$ has been applied. We see that there is a phase twist propagating along the first arm of the paths, and the phase twist due to the pulse sent at $t - 4\tau_F$ is still propagating through the second, third etc. paths. Similarly the phase twist sent by the pulse at $t - 8\tau_F$ is still propagating

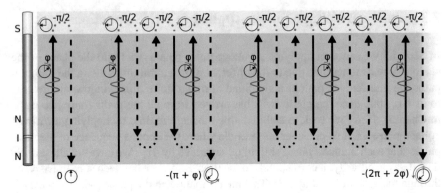

Fig. 7.5 Snapshot of the effect of a voltage pulse train with period $T = 4\tau_F$ on the relative phases of the paths through the NINS junction. At $E = 0$ the only phase picked up is due to Andreev reflection ($-\pi/2$ phase shift) and the voltage pulse (φ phase shift). We define the first path to have a phase 0; longer paths have a greater number of factors of $-\pi/2$ due to the greater number of Andreev reflections, but also at any given time fewer voltage-pulse-induced phase shifts have arrived compared to the shorter paths, which gives path n a phase shift of $-\varphi$ compared to path $n-1$

through the third, fourth etc. paths. When the period of the pulse train exactly matches the delay between different trajectories, $T = 4\tau_F$, the phase $\varphi(t + 4m\tau_F)$ is simply given by $\varphi(t + 4m\tau_F) = 2\pi\bar{n}m$. As a result, $\Phi_m(t, E) = m\Phi(E) + 2\pi\bar{n}m$ and (7.16) takes the form of a geometric series. We see that the application of such a train of pulses just *shifts* the resonance by $2\pi\bar{n}$, as illustrated in Fig. 7.6b.

For the case of an *alternating* train of pulses the situation is a little more complicated; now, the positive pulse induces a phase shift of $2\pi\bar{n}$ and the negative pulse induces a phase shift of $-2\pi\bar{n}$. If we tune the pulse train such that there is $4\tau_F$ between successive positive/negative pulses (so that the period is $8\tau_F$), then *alternate* paths pick up a phase of $2\pi\bar{n}$. We see in Fig. 7.6c that the effect of an alternating pulse train has a qualitatively different effect; now the positions of the resonances are fixed and changing $\bar{n}$ just changes the relative amplitude of the peaks. Note that for $\bar{n} = 0.5$ the effect of the upright and alternating pulse trains is identical.

Now that we have some intuition for what is going on we shall proceed with the generic case where T is not a multiple of $4\tau_F$. We expand $\varphi(t)$ in terms of its frequency components as:

$$e^{i\varphi(t)} = e^{i\varphi(T)t/T} \sum_{p=-\infty}^{+\infty} c_p e^{ip\omega t}, \tag{7.18}$$

where $\omega = 2\pi/T$. The extra factor in front of the usual Fourier series takes care of the case when the average of $V_P(t)$ over one period is different from zero (which is the case for the upright pulse train), and hence $\varphi(t)$ is not periodic. By inserting

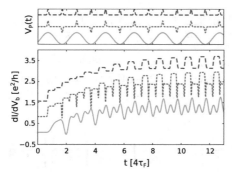

(a) Time-resolved differential conductance (lower figure) at the normal contact of the NINS junction for three voltage pulse trains (upper figure), calculated from numerical simulations using eq. (7.9).

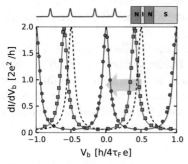

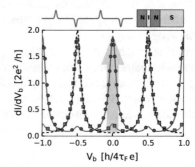

(b) Time averaged differential conductance for a train of upright pulses.

(c) Time averaged differential conductance for a train of alternating pulses.

Fig. 7.6 Numerical simulation of the differential conductance in an NINS junction in the presence of a train of voltage pulses. In (**b**) and (**c**) the *solid curves* were calculated (semi-)analytically using Eq. (7.19) and symbols were calculated from numerical simulations using Eq. (7.10). *Squares* correspond to $\bar{n} = 0.1$, circles correspond to $\bar{n} = 0.5$ and *dashed lines* correspond to the d.c. case. The *upright* pulse train used in (**b**) has period $T = 4\tau_F$ and the alternating train in (**c**) has period $T = 8\tau_F$

Eq. (7.18) into Eq. (7.16) and keeping only the d.c. component (i.e. the time independent part), we obtain

$$\left\langle \frac{\partial I(t)}{\partial V_b} \right\rangle_T = \frac{4e^2|d|^4}{h} \sum_{p=-\infty}^{+\infty} \left| \frac{c_p}{1 - |r|^2 e^{i\Phi(eV_b) + 4i\omega\tau_F[\varphi(T)/2\pi + p]}} \right|^2, \tag{7.19}$$

where we have explicitly replaced the energy E with eV_b. As before, the appearance of resonances corresponds to minimising the denominator of Eq. (7.19), i.e. the phase factor should be a multiple of 2π. This translates into the following resonance condition:

$$- 2 \arccos \left(\frac{eV_b}{\Delta_0} \right) + \frac{4eV_b\tau_F}{\hbar} + 4\omega\tau_F \left[p + \frac{\varphi(T)}{2\pi} \right] = 2\pi q \quad p, q \in \mathbb{Z}. \quad (7.20)$$

If we are only concerned with resonances far below the gap, such that $eV_b \ll \Delta_0$, we can expand $\arccos(eV_b/\Delta_0) \approx \pi/2 - eV_b/\Delta_0$, however in the long junction limit (that we shall consider now) where $\Delta_0 \gg \hbar/\tau_F$ the term linear in eV_b/Δ_0 can be neglected and we arrive at

$$\frac{eV_b}{h} + \frac{\omega}{2\pi} \left[p + \frac{\varphi(T)}{2\pi} \right] = \frac{1}{4\tau_F} \left[q + \frac{1}{2} \right] \quad p, q \in \mathbb{Z}. \quad (7.21)$$

In the opposite (short junction) limit the above analysis is not complete, as the pulses must be shorter (or of the same order, as we shall see in the following paragraphs) than τ_F, if $\hbar/\tau_F \sim \Delta_0$ then the pulses will necessarily excite states at energies above the superconducting gap. This quasiparticle current has not been taken into account in the above analysis.

An illustration of the resonance condition is shown rather beautifully in Fig. 7.7, where the contributions from various terms of order p from Eq. (7.19) are shown (Fig. 7.7a–d) as a function of the pulse frequency and bias voltage for the case of a sinusoidal $V_P(t)$. We can consider the sinusoid as a deformation of the alternating pulse train previously considered. We note, however, that the even though the duration of a single "pulse" ($4\tau_F$) is of the same order of magnitude as the frequency of the

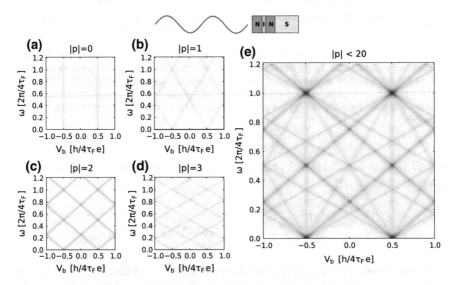

Fig. 7.7 Contribution to the differential conductance in the NINS junction in the presence of a sinusoidal voltage "pulse train" of period $T = 8\tau_F$ from different orders p of the sum Eq. (7.19). **a–d** The first 4 orders in the sum. **e** The sum of the first 21 orders in the sum; higher order terms are negligibly small

train $(8\tau_F)$, the above-described interference effect is not washed out. Although Fig. 7.7 was obtained by numerically evaluating Eq. (7.19), we also performed full tight-binding simulations using the model Eq. (7.6) and the agreement is essentially perfect. This can also be seen in the perfect agreement between the symbols and lines in Fig. 7.6b and c.

When $\omega = 0$ we clearly recover the d.c. resonant peaks shown in Fig. 7.4, and at finite frequency these "split" only to come back together at $\omega = 2\pi/4\tau_F$. This "re-emergence" of the d.c. interference pattern is actually rather trivial, as when $T = 4\tau_F$ the extra relative phase between subsequent paths is always zero with respect to the d.c. case (for the case of alternating/sinusoidal pulses, where a net phase of zero is picked up over one period). For the case of $T = 8\tau_F$ *alternating* paths pick up a phase $2\pi\bar{n}$ with respect to one another, which is what gives rise to the peak at zero bias.

7.4 Majorana States in NINS Junctions

Let us switch gears for a moment and come back to the possible applications of the above-described technique. We said in the introduction to this chapter that we wanted to use our technique to manipulate Majorana resonances that have been theorised to exist in NINS junctions. We will essentially see that Majorana resonances in NINS junctions can be understood using the same resonant Andreev reflection mechanism described in the preceding sections.

In order to appreciate why this is important, and why it is indeed possible, it will be necessary to first describe what these Majorana *are* and why the search for them is the subject of such an intense theoretical as well as experimental interest [1, 8–13]. While it is not the goal of this section to give an in-depth review (this has been done exceptionally well elsewhere [14–19]), it is naturally necessary to give an introduction to the topic.

7.4.1 Introduction to Majorana States

In 1937 Ettore Majorana proposed a new formulation of the Dirac equation that predicted the existence of particles that are their own antiparticle [20], which have come to be known as Majorana particles (henceforth shortened to "Majoranas"). Although no definitive realisation of Majoranas as an elementary particle has so far been observed, it has been known for some time that such objects could exist as quasiparticle excitations in superconductors [21–24]. Although on an aesthetic level the discovery of such quasiparticles in condensed matter systems would represent a triumph of modern physics and engineering in itself, it can be argued that this is not the practical motivation for the search; the real aim is to develop an additional platform for quantum computing that is more robust to the dreaded decoherence that plagues existing approaches.

To understand this let us first see how it is possible to realise Majoranas in systems exhibiting superconductivity. The seminal paper by Kitaev [24] proposed a simple theoretical model that contains the essential ingredients required for Majoranas. The Hamiltonian for the model takes the following form:

$$\hat{H} = \sum_j -\gamma(\hat{a}_j^\dagger \hat{a}_{j+1} + \hat{a}_{j+1}^\dagger \hat{a}_j) - \mu \hat{a}_j^\dagger \hat{a}_j + \Delta_0 \hat{a}_j^\dagger \hat{a}_{j+1}^\dagger + \Delta_0^* \hat{a}_{j+1} \hat{a}_j, \qquad (7.22)$$

where γ is a hopping amplitude, μ is a chemical potential, Δ_0 is the induced superconducting gap and $\hat{a}_j^\dagger$ ($\hat{a}_j$) are fermionic creation (destruction) operators for a state on site i. The presence of anomalous terms $\hat{a}_{j+1} \hat{a}_j$ looks very much like the models of superconductors we have seen thus far, except that the superconducting terms couple *different* sites, and there is only a single spin species involved. One can then define *Majorana operators*

$$\hat{\zeta}_j^A = \frac{\hat{a}_j + \hat{a}_j^\dagger}{\sqrt{2}}$$
$$\hat{\zeta}_j^B = \frac{\hat{a}_j - \hat{a}_j^\dagger}{i\sqrt{2}}, \qquad (7.23)$$

so called because they have the property that $[\hat{\zeta}_j^A]^\dagger = \hat{\zeta}_j^A$ and $[\hat{\zeta}_j^B]^\dagger = \hat{\zeta}_j^B$; creating such an excitation is the same thing as destroying it. Rewriting Eq. (7.22) in terms of these new operators we get

$$\hat{H} = \frac{i}{2} \sum_j -\mu \hat{\zeta}_j^A \hat{\zeta}_j^B + (\gamma + \Delta_0) \hat{\zeta}_j^B \hat{\zeta}_{j+1}^A + (-\gamma + \Delta_0) \hat{\zeta}_j^A \hat{\zeta}_{j+1}^B. \qquad (7.24)$$

We can now take two different parameter regimes for the model:

- the trivial one where $\gamma = \Delta_0 = 0$, $\mu < 0$, and $\hat{H} = \frac{-i\mu}{2} \sum_j \hat{\zeta}_j^A \hat{\zeta}_j^B$,
- and the nontrivial one where $\Delta_0 = \gamma > 0$, $\mu = 0$, and $\hat{H} = i\gamma \sum_j \hat{\zeta}_j^B \hat{\zeta}_{j+1}^A$.

The first case is trivial in the sense that it corresponds to a chain of sites with no coupling between them. We see that in this case the Hamiltonian is just a sum of terms that "pair" the two Majorana operators on each site; the terms $i\hat{\zeta}_j^a \hat{\zeta}_j^B = \hat{a}_j^\dagger \hat{a}_j - 1/2$ count the original fermions on site j. In the nontrivial case we see that now Majorana from *neighbouring* sites pair up; this pairing is illustrated in Fig. 7.8. For the case of a finite chain of N sites, we thus see that the Majoranas $\hat{\zeta}_1^A$ and $\hat{\zeta}_N^B$ do not appear in the Hamiltonian at all. The consequence of this is that there is a very non local fermionic state formed of the superposition of these two Majorana states, $\hat{f}^\dagger = \left(\hat{\zeta}_1^A - i\hat{\zeta}_N^B\right)/\sqrt{2}$, that has *zero energy*. The ground state is therefore twofold degenerate: If $|0\rangle$ is a ground state satisfying $\hat{f}|0\rangle = 0$, then $|1\rangle = \hat{f}^\dagger|0\rangle$ is also a

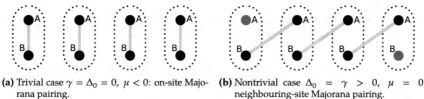

(a) Trivial case $\gamma = \Delta_0 = 0$, $\mu < 0$: on-site Majorana pairing.

(b) Nontrivial case $\Delta_0 = \gamma > 0$, $\mu = 0$: neighbouring-site Majorana pairing.

Fig. 7.8 Illustration of the two Majorana pairing regimes present in different parameter ranges of the model Eq. (7.22) for a four-site chain. The *circles* labelled A and B represent the Majoranas (see main text): They are grouped by site (*the dotted outline*) and the *grey lines* represent hopping terms in the Hamiltonian Eq. (7.24). **b** Shows the two unpaired Majoranas that do not appear in the Hamiltonian

ground state. In addition these two states are separated from the rest of the states by an energy gap Δ_0, as can be seen from the above model.

This non local state formed from two Majoranas is what the community is searching for. Such a state should be robust to a whole host of decoherence mechanisms due to the fact that it is non local and is separated from the rest of the states by an energy gap. Much has been said in the literature about this so-called "topological" protection [25, 26]. The end goal is to use such states to encode quantum information and realise quantum logic operations by *braiding* the constituent Majorana of several such delocalised states around one another [25, 26]. A description of how to perform computational operations with these objects falls well outside the scope of this thesis, however; we shall content ourselves with the question of how to realise such states experimentally and how to detect them.

7.4.2 Realisation of Majorana States In Nanowires

While the Kitaev model is certainly simple it is not immediately clear how one could realise such a Hamiltonian experimentally. Several proposals were made for how to generate the necessary theoretical components using experimentally available ingredients, in particular: coupling regular superconductors to two-dimensional electron gases in the fractional quantum Hall regime [27, 28], and coupling regular superconductors to semiconductor nanowires or thin films with strong spin-orbit coupling while applying magnetic field [29–33]. We shall concentrate on this latter proposal, specifically that of Ref. [29] on which subsequent experiments were based [1].

The model proposed in Ref. [29] is for a nanowire with Rashba spin-orbit coupling and with a magnetic field oriented along the axis of the nanowire, coupled to a regular s-wave superconductor. If we treat the nanowire as having an effective superconducting gap Δ_0, induced by proximity effect from the coupled superconductor [33, 34], The Hamiltonian for this system is written as

$$\hat{H} = \int_0^L dx\, \hat{\Psi}^\dagger(x) \left[\left(\frac{-\hbar^2}{2m^*} \frac{\partial^2}{\partial x^2} - i\hbar\alpha\sigma_1 \frac{\partial}{\partial x} - E_F \right) \tau_3 + E_Z\sigma_3 + \Delta_0\tau_1 \right] \hat{\Psi}(x)$$

(7.25)

where L is the length of the nanowire, α quantifies the spin-orbit coupling, E_Z is the Zeeman energy, $\hat{\Psi}(x)$ is defined by Eq. (7.2), and σ_n are defined analogously to the τ_n in Eq. (7.4):

$$\sigma_1 = \mathbb{1}_2 \otimes \begin{pmatrix} 0 & 1 \\ 1 & 0 \end{pmatrix}, \quad \sigma_2 = \mathbb{1}_2 \otimes \begin{pmatrix} 0 & -i \\ i & 0 \end{pmatrix}, \quad \sigma_3 = \mathbb{1}_2 \otimes \begin{pmatrix} 1 & 0 \\ 0 & -1 \end{pmatrix}$$

(7.26)

where $\mathbb{1}_2$ denotes a 2×2 identity matrix and $\otimes$ is the Kronecker product. When the parameters of this model are tuned such that $E_Z^2 > E_F^2 + \Delta_0^2$ then this model exhibits a "topological phase" that hosts Majorana zero modes at the ends of the nanowire. Figure 7.9a shows a numerical calculation of spectrum of this model discretized onto a lattice. We see that as E_Z is increased the gap closes at $E_Z = \sqrt{E_F^2 + \Delta_0^2}$ and then reopens, however the lowest eigenenergy is now pinned to $E = 0$. Figure 7.9b shows the square modulus of the wavefunction for the lowest energy state in the "trivial" ($E_Z < \sqrt{E_F^2 + \Delta_0^2}$) and "topological" ($E_Z > \sqrt{E_F^2 + \Delta_0^2}$) phases. In the latter case we can clearly see its delocalised nature; this is the Majorana state.

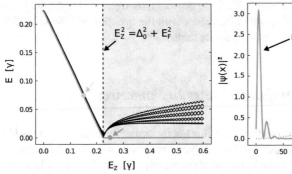
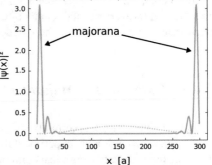

(a) Ten lowest eigenenergies as a function of Zeeman energy E_Z. The shaded region where $E_Z^2 > \Delta_0^2 + E_F^2$ corresponds to the topological regime. We can observe the lowest energy eigenstate, which sticks to zero energy in the topological regime; this is the Majorana state.

(b) Absolute square of the electron part of the wavefunction (summed over the spin degree of freedom) for the smallest energy eigenvalue. The solid line is for $E_Z = 0.25\gamma$ (indicated by a solid arrow in (a)), and the dotted line is for $E_Z = 0.15\gamma$ (indicated by a dotted arrow in (a)).

Fig. 7.9 d.c. Numerical simulations of the model Eq. (7.25) discretised with spacing a onto a chain 300 sites long. The parameters used were $\Delta_0 = 0.2\gamma$, $\beta = 0.05\gamma$, and $E_F = 0.1\gamma$, where $\gamma = \hbar^2/2m^*a^2$ and $\beta = \hbar\alpha/a$. The energy spectrum (**a**) and zero-mode wavefunction (**b**) were obtained by directly diagonalising the resulting 300×300 matrix

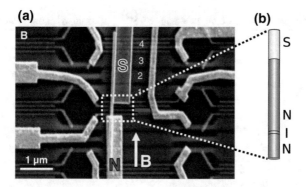

Fig. 7.10 Realisation of a setup for producing Majorana quasiparticles. **a** Scanning electron microscope image of a normal-insulator-normal-superconductor (NINS) junction that purportedly supports Majoranas (from Ref. [1], reprinted with permission from AAAS). **b** Identification of our model with the experimental setup. The insulating barrier could correspond either to one of the gates (barely visible between the N and S regions in (a)), or the junction between the nanowire and the Normal lead

Experimental Realisation

The next question to ask is how one can experimentally *observe* such objects; in the above model the nanowire is isolated and the Majoranas are true bound states. If one weakly couples a regular conductor to the end of the nanowire, however, the bound state will couple to the continuum of states in the regular conductor and become a *resonance*, with a corresponding distinctive peak in the differential conductance centred at the energy of the bound state, that is, at the Fermi energy: a zero bias peak [35–37]. This proposal was later realised experimentally in the seminal work of Mourik et al. [1], where they claimed to have generated these Majorana quasiparticles. Their setup is shown in Fig. 7.10: They have an indium antimonide nanowire coupled to a superconducting niobium titanium nitride contact (S) and normal gold contact (N), and a set of metallic gates electrostatically coupled to the nanowire provide an insulating barrier. In their experiment they measured the differential conductance of this NINS junction in the parameter regime where one would expect to find Majoranas, and observed a peak at zero bias.

7.4.3 Scattering Perspective on Majorana Resonances

Let us put the Majorana interpretation to one side for a moment and just consider at face value the model presented in Eq. (7.25). Figure 7.11 shows the band structure for the leads as successive ingredients are added. Initially the two spin bands are degenerate, and this degeneracy is then split when the Rashba term is added; there are still *four* states available at the Fermi energy ($E = 0$ in Fig. 7.11). However, when the Zeeman term is added the spin bands hybridise and a gap opens near $k = 0$. At certain

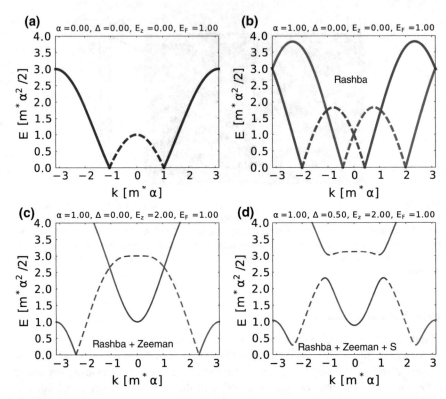

Fig. 7.11 Bandstructures in the excitation picture for the model Eq. (7.25) as additional ingredients are added. Solid lines denote electron states, while *dashed lines* denote hole states. *Red* colouration denotes spin up, while *blue* denotes spin down. **a** Simple spin-degenerate system. The spin bands are degenerate; there are 4 states available at the Fermi energy. **b** System with Rashba coupling only. The spin bands have been shifted but there are still 4 states available at the Fermi energy. **c** System with Rashba and Zeeman coupling. There are now only 2 states available at the Fermi energy, and the spin is locked to the momentum. **d** System with Rashba, Zeeman coupling and superconductivity

energies there are therefore only *two* states available, which are counterpropagating and have opposite spin (the case of Fig. 7.11c). This "spin-momentum locking" has important consequences for transport in the full NINS junction at these energies; notably, when the electrons/holes are reflected at the normal barrier they *must* undergo a spin flip. Figure 7.12 illustrates the differences between the case with/without spin-momentum locking. We see that the amplitude $\rho_{e\uparrow,h\downarrow}$ in the second (and subsequent) paths becomes $\rho_{e\downarrow,h\uparrow}$: We change spin sectors, which means that the second path picks up an *extra minus sign* with respect to the case without spin-momentum locking, because $\rho_{e\uparrow,h\downarrow} = -\rho_{e\downarrow,h\uparrow}$. The result of this extra minus sign is to shift the positions of the resonances, such that there is a resonance at $E = 0$ (the Fermi energy). We can thus rewrite Eq. (7.21) as

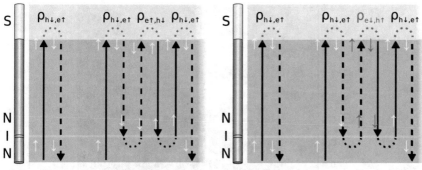

(a) In absence of spin-momentum locking. (b) In presence of spin-momentum locking.

Fig. 7.12 First couple of paths through the NINS junction that contribute to the amplitude $r_{h\downarrow,e\uparrow}$, with spin orientation explicitly marked. $\rho_{h\downarrow,e\uparrow}$ is the amplitude for Andreev reflection from a spin up electron to a spin down hole. In (**b**) there is spin-momentum locking; the difference with (**a**) is highlighted in *red*

$$\frac{eV_b}{h} + \frac{\omega}{2\pi}(p + \varphi(T)/2\pi) = \frac{1}{4\tau_F}\left[q - \frac{1}{2}(Z_2 - 1)\right] \quad p, q \in \mathbb{Z} \qquad (7.27)$$

where Z_2 is 0 in the "normal" case, and 1 when there is spin-momentum locking. We can also write

$$\left\langle \frac{\partial I(t)}{\partial V_b} \right\rangle_T = \frac{4e^2|d|^2}{h(1 + Z_2)} \left| r_{h\downarrow,e\uparrow}(t, E) \right|^2 \qquad (7.28)$$

where $r_{h\downarrow,e\uparrow}(t, E)$ is given by Eq. (7.16) with $\Phi(E)$ now given by

$$\Phi(E) = -2\arccos\left(\frac{E}{\Delta_0}\right) + \frac{4E\tau_F}{\hbar} + \pi Z_2. \qquad (7.29)$$

We see that there will be a resonance at $E = 0$ which gives rise to a $2e^2/h$ conductance peak at zero bias (the magnitude is halved with respect to the "normal" case because there are now only half the number of states available at the Fermi energy). This *is* the resonant Andreev reflection from the Majorana state discussed previously [6, 35].

7.5 Manipulating Majorana Resonances with Voltage Pulses

Now with our understanding of how to manipulate Andreev resonances using trains of voltage pulses, combined with the view of Majorana resonances as Andreev resonances (just with an "extra minus sign"), we now arrive at an almost banal conclusion: we can manipulate Majorana resonances using trains of voltage pulses.

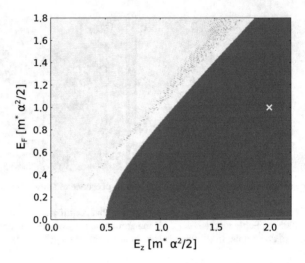

Fig. 7.13 Zero-bias differential conductance for the NINS junction as a function of the Zeeman (E_Z) and Fermi (E_F) energies, with $\alpha = 1$, $\Delta = 0.5$ and $V_T = 5.0$ (transmission $D = 0.17$). The colour scale goes between 0 (*white*) and $2e^2/h$ (*dark grey*). The continuous phase where $\partial I/\partial V_b = 2e^2/h$ in the $\sqrt{E_Z^2 - \Delta^2} > E_F$ sector corresponds to the "topological" regime where there is a Majorana resonance. The isolated points of high conductance around $E_Z = E_F$ correspond to regular Andreev resonances crossing zero bias. The white cross indicates where in this parameter space the time-dependent numerical simulations were carried out

We take the following model

$$
\mathbf{H}_{\mathrm{BdG}}(x, t) = \left[\frac{-\hbar^2}{2m^*} \frac{\partial}{\partial x^2} - i\hbar\alpha\sigma_1 \frac{\partial}{\partial x} + qV_T\delta(x) + qV_P(t)\Theta(-x) - E_F \right] \tau_3
$$
$$
+ E_Z\sigma_3 + \Delta_0\Theta(y - L)\tau_1
$$

$$
(7.30)
$$

and discretise onto a lattice as before to obtain the a tight-binding Hamiltonian with matrix elements $\mathbf{H}_{i,j}$ given by

$$
\mathbf{H}_{j,j} = [2\gamma - E_F + qV_T\delta_{0,j}]\tau_3 + E_Z\sigma_3 + \Delta_0\Theta_{j,l_j}\tau_1
$$
$$
\mathbf{H}_{j,j+1} = -\gamma e^{i\varphi_P(t)\delta_{-1,j}\tau_3}\tau_3 - i\hbar\alpha\sigma_1\tau_3
$$

$$
(7.31)
$$

Figure 7.13 shows the zero-bias conductance (the colour scale) as a function of the Fermi and Zeeman energies (E_F and E_Z respectively), calculated numerically from the above model. We clearly see a phase transition where a zero bias (Majorana) peak appears when the condition $E_F^2 < E_Z^2 - \Delta_0^2$ is satisfied. In what follows we placed ourselves at the white cross in the above phase diagram, so that the system exhibits Majoranas in the absence of any time dependence. Figure 7.14a shows the conductance as a function of bias voltage when a train of alternating pulses is applied. We see that the application of the voltage pulses *reduces* the zero-bias peak, analogous

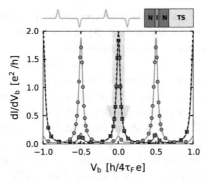

(a) Differential conductance in d.c. (dashed curve), and in the presence of a train of voltage pulses with period $T = 8\tau_F$ and $\bar{n} = 0.1$ (solid curve and square symbols) and $\bar{n} = 0.5$ (solid curve and circle symbols). Solid curves were calculated using eq. (7.28) and the symbols were calculated numerically.

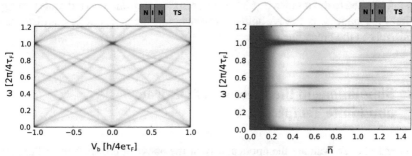

(b) Differential conductance in the presence of a sinusoidal voltage with $\bar{n} = 0.5$ as a function of frequency and bias voltage.

(c) Differential conductance at zero bias in the presence of a sinusoidal voltage pulse as a function of frequency and $\bar{n}$.

Fig. 7.14 Differential conductance in the presence of a train of voltage pulses for a NINS junction that displays a Majorana resonance in d.c.

to the case of Fig. 7.6c (without Majoranas), where the same voltage pulse train created a zero-bias peak. Figure 7.14b is the analogue of Fig. 7.7e, and just illustrates that the resonances are shifted with respect to the "normal" case (as implied by Eq. (7.27)). Figure 7.14c shows the differential conductance at zero bias in the presence of a sinusoidal $V_P(t)$, as a function of frequency and the phase picked up under each "lobe" of the sinusoid. We can see that arbitrary frequencies/phases will destroy the zero-bias peak, which is unsurprising as in such a case subsequent paths will have an arbitrary phase relationship that will result in destructive interference. Strikingly we see the *re-emergence* of the zero-bias peak when the period is an integer multiple of $4\tau_F$ and at distinct values of $\bar{n}$. The position in frequency of these re-emergences is simple to understand, as these are the only frequencies that will give a fixed-phase relationship between the different paths.

Despite their simplicity, Fig. 7.14b, c provide very strong signatures for the Majorana mode. This is promising, as it indicates that this scheme could be used as an extra verification for the Majorana interpretation of current experiments. There was originally some controversy around the initial findings of Ref. [1], as there could in principle be a number of explanations for a zero-bias peak that did not involve Majorana physics at all. Among the proposed alternative explanations for the zero-bias peak were: Kondo resonance [38–41], interplay between multiple sub-bands [42–44], weak antilocalisation [45], and the effect of boundary conditions [46–48]. Although it is true that the present proposal does not address these concerns directly, it effectively adds two more parameters (pulse height, related to $\bar{n}$, and train period T) with which to probe the system, and so can only serve to add more information about what physics is actually at play. As the intricate way in which the conductance is altered is intrinsically linked to coherent Andreev reflection processes, it is difficult to see how zero bias peaks due to the alternative mechanisms mentioned above could be affected in the same way if their physics is not at all related. We would however make one concession and note that the technique presented here could probably not be used to distinguish true Majorana resonances from low (but finite) energy Andreev resonances, as they are both due to essentially the same physics, as discussed extensively in the preceding pages.

7.6 Simulations in the Presence of Disorder and Finite Temperature

In order to better evaluate the applicability of the above approach to a real experimental system it is necessary to include a few more ingredients into the model. For example, in the above we worked in a long junction regime where there are a number of resonances below the (proximity induced) superconducting gap, whereas present experiments are typically in the short junction regime where the mean level spacing between the resonances is of the order of/much larger than the gap. Specifically a junction length of 200 nm corresponds (using a Fermi velocity of 10^4–10^5 m s^{-1}) to a mean level spacing of 50–500 μeV, i.e. of the order of the superconducting gap, which was 250 μeV in the experimental setup of Ref. [1]. In addition the nanowires used typically have a mean free path of roughly the same length as the junction itself (in Ref. [1] it was measured to be 300 nm), in contrast to the perfectly clean case studied in the preceding sections. As our mechanism relies heavily on the interference between paths with well-defined lengths, it is not obvious that the addition of disorder (which effectively adds a greater number of possible paths due to backscattering) will not destroy the effect completely. Finally the experiments are done at finite temperature, and so we should also look at the effect of thermal broadening on our mechanism.

The model parameters were tuned to be similar those in the experimental setup of Ref. [1]. Specifically, we chose $\Delta = 250\,\mu$eV, $\alpha = 20$ meV nm, $E_F = 0$, and a magnetic field of 0.6 T to place us firmly in the topological regime. We used a discretisation step of 1 nm, which at the relevant energy scales Δ gives

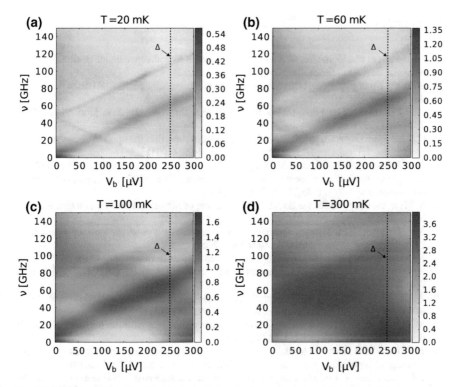

Fig. 7.15 Differential conductance as a function of bias voltage and voltage pulse frequency (for a sinusoisal voltage "pulse train", for different temperatures, of an NINS junction that displays a Majorana resonance in d.c. The colour bars give the colour scale for each plot in units of e^2/h

a band structure that is negligibly different from the continuum limit. Using the above microscopic parameters the Fermi velocity was calculated numerically to be $6.08 \times 10^4\,\mathrm{ms}^{-1}$ which in principle—for the 100 nm junctions studied in our simulations—gives a mean-level spacing of 630 μeV. Finally the disorder used gives a mean free path of 87 nm, which is of the order of the junction length and places us in the quasi-ballistic regime, similar to the experiments.

Figure 7.15 shows a repeat of Fig. 7.14b (i.e. sinusoidal "pulse" train) using the model described in the preceding paragraph. The different subplots show the results at different temperatures. Unsurprisingly the addition of disorder destroys some of the finer detail, and the presence of only a single resonance (the Majorana resonance) below the gap means that there is a less rich sub-gap structure. Nevertheless, the application of the time-dependent bias still does shift the resonance from zero bias; the key feature of the signature remains. It should be noted that the addition of the voltage pulses does not add significant noise to the d.c. signal; if a resonance is well-resolved and visible in a d.c. experiment, then it should also be visible in the presence of the voltage pulses.

References

1. V. Mourik et al., Signatures of Majorana fermions in hybrid superconductor-semiconductor nanowire devices. Science **336**(6084), 1003–1007 (2012)
2. B. Gaury, Emerging concepts in time-resolved quantum nanoelectronics, PhD thesis, UniversitT de Grenoble, Oct 2014
3. C.W.J. Beenakker, Random-matrix theory of quantum transport. Rev. Mod. Phys. **69**(3), 731–808 (1997)
4. C.J. Lambert, Generalized Landauer formulae for quasi-particle transport in disordered superconductors. J. Phys. Condens. Matter **3**(34), 6579–6587 (1991)
5. G.E. Blonder, M. Tinkham, T.M. Klapwijk, Transition from metallic to tunneling regimes in superconducting microconstrictions: excess current, charge imbalance, and supercurrent conversion. Phys. Rev. B **25**(7), 4515–4532 (1982)
6. S. Mi et al., Proposal for the detection and braiding of Majorana fermions in a quantum spin hall insulator. Phys. Rev. B **87**(24), 241405 (2013)
7. Y.V. Nazarov, Y.M. Blanter, Quantum transport: introduction to nanoscience (Cambridge University Press, Cambridge, UK ; New York, 2009)
8. H.O.H. Churchill et al., Superconductor-nanowire devices from tunneling to the multichannel regime: zero-bias oscillations and magnetoconductance crossover. Phys. Rev. B **87**(24), 241401 (2013)
9. A. Das et al., Zero-bias peaks and splitting in an Al-InAs nanowire topological superconductor as a signature of Majorana fermions. Nat Phys **8**(12), 887–895 (2012)
10. M.T. Deng et al., Anomalous zero-bias conductance peak in a Nb-InSb nanowire-Nb hybrid device. Nano Lett. **12**(12), 6414–6419 (2012)
11. M.T. Deng et al., Parity independence of the zero-bias conductance peak in a nanowire based topological superconductor-quantum dot hybrid device. Sci. Rep. **4**, 7261 (2014)
12. L.P. Rokhinson, X. Liu, J.K. Furdyna, The fractional a.c. Josephson effect in a semiconductor-superconductor nanowire as a signature of Majorana particles. Nat Phys **8**(11), 795–799 (2012)
13. A.D.K. Finck et al., Anomalous modulation of a zero-bias peak in a hybrid nanowire-superconductor device. Phys. Rev. Lett. **110**(12), 126406 (2013)
14. C.W.J. Beenakker, Search for Majorana fermions in superconductors. Ann. Rev. Condens. Matter Phys. **4**(1), 113–136 (2013). arXiv: 1112.1950
15. J. Alicea, New directions in the pursuit of Majorana fermions in solid state systems. Rep. Prog. Phys. **75**(7), 076501 (2012)
16. C.W.J. Beenakker, Random-matrix theory of Majorana fermions and topological superconductors. Rev. Mod. Phys. **87**(3), 1037–1066 (2015)
17. M. Leijnse, K. Flensberg, Introduction to topological superconductivity and Majorana fermions. Semicond. Sci. Technol. **27**(12), 124003 (2012)
18. T.D. Stanescu, S. Tewari, Majorana fermions in semiconductor nanowires: fundamentals, modeling, and experiment. J. Phys. Condens. Matter **25**(23), 233201 (2013)
19. S.D. Sarma, M. Freedman, C. Nayak, Majorana zero modes and topological quantum computation. NPJ Quantum Information 1 (Oct. 2015), p. 15001
20. E. Majorana, Teoria simmetrica dell'elettrone e del positrone. Nuovo Cim **14**(4), 171–184 (1937)
21. G.E. Volovik, Fermion zero modes on vortices in chiral superconductors. J. Exp. Theor. Phys. Lett. **70**(9), 609–614 (1999)
22. T. Senthil, M.P.A. Fisher, Quasiparticle localization in superconductors with spinorbit scattering. Phys. Rev. B **61**(14), 9690–9698 (2000)
23. N. Read, D. Green, Paired states of fermions in two dimensions with breaking of parity and time-reversal symmetries and the fractional quantum Hall effect. Phys. Rev. B **61**(15), 10267–10297 (2000)
24. A.Y. Kitaev. Unpaired Majorana fermions in quantum wires. Phys.-Usp. **44**(10S), 131 (2001)
25. A.Y. Kitaev, Fault-tolerant quantum computation by anyons. Ann. Phys. **303**(1), 2–30 (2003)

26. C. Nayak et al., Non-Abelian anyons and topological quantum computation. Rev. Mod. Phys. **80**(3), 1083–1159 (2008)
27. G. Moore, N. Read, Nonabelions in the fractional quantum hall effect. Nucl. Phys. B **360**(2), 362–396 (1991)
28. S.D. Sarma, M. Freedman, C. Nayak. Topologically protected qubits from a possible Non-Abelian fractional quantum hall state. Phys. Rev. Lett. **94**(16), 166802 (2005)
29. Y. Oreg, G. Refael, F. von Oppen, Helical liquids and Majorana bound states in quantum wires. Phys. Rev. Lett. **105**(17), 177002 (2010)
30. J. Alicea, Majorana fermions in a tunable semiconductor device. Phys. Rev. B **81**(12), 125318 (2010)
31. R.M. Lutchyn, J.D. Sau, S.D. Sarma, Majorana fermions and a topological phase transition in semiconductor-superconductor heterostructures. Phys. Rev. Lett. **105**(7), 077001 (2010)
32. J.D. Sau, et al., Generic new platform for topological quantum computation using semiconductor heterostructures. Phys. Rev. Lett. **104**(4), 040502 (2010)
33. C.L. Kane, Superconducting proximity effect and Majorana fermions at the surface of a topological insulator. Phys. Rev. Lett. **100**(9), 096407 (2008)
34. T.M. Klapwijk, Proximity effect from an Andreev perspective. J. Supercond. **17**(5), 593–611 (2004)
35. K.T. Law, P.A. Lee, T.K. Ng, Majorana fermion induced resonant Andreev reflection. Phys. Rev. Lett. **103**(23), 237001 (2009)
36. K. Flensberg, Tunneling characteristics of a chain of Majorana bound states. Phys. Rev. B **82**(18), 180516 (2010)
37. J.D. Sau et al., Phys. Rev. B **82**(21), 214509 (2010)
38. W. Chang et al., Tunneling spectroscopy of quasiparticle bound states in a spinful Josephson junction. Phys. Rev. Lett. **110**(21), 217005 (2013)
39. M. Cheng et al. Interplay between Kondo and Majorana interactions in quantum dots. Phys. Rev. X **4**(3), 031051 (2014)
40. E.J.H. Lee et al., Zero-bias anomaly in a nanowire quantum dot coupled to superconductors. Phys. Rev. Lett. **109**(18), 186802 (2012)
41. Rok Žitko et al., Shiba states and zero-bias anomalies in the hybrid normal-superconductor Anderson model. Phys. Rev. B **91**(4), 045441 (2015)
42. G. Kells, D. Meidan, P.W. Brouwer, Low-energy subgap states in multichannel p-wave superconducting wires. Phys. Rev. B **85**(6), 060507 (2012)
43. J. Liu et al., Zero-bias peaks in the tunneling conductance of spin-orbit-coupled superconducting wires with and without Majorana end-states. Phys. Rev. Lett. **109**(26), 267002 (2012)
44. R.M. Lutchyn, T.D. Stanescu, S.D. Sarma, Search for Majorana fermions in multiband semiconducting nanowires. Phys. Rev. Lett. **106**(12), 127001 (2011)
45. D.I. Pikulin et al., A zero-voltage conductance peak from weak antilocalization in a Majorana nanowire. New J. Phys. **14**(12), 125011 (2012)
46. G. Kells, D. Meidan, P.W. Brouwer, Near-zero-energy end states in topologically trivial spinorbit coupled superconducting nanowires with a smooth confinement. Phys. Rev. B **86**(10), 100503 (2012)
47. M.-T. Rieder et al., Endstates in multichannel spinless p-wave superconducting wires. Phys. Rev. B **86**(12), 125423 (2012)
48. T.D. Stanescu, T. Sumanta, Nonlocality of zero-bias anomalies in the topologically trivial phase of Majorana wires. Phys. Rev. B **89**(22), 220507 (2014)

Chapter 8
Conclusion

Experiments that probe the internal dynamics of quantum devices are beginning to be realised in the laboratory, opening the way for a whole host of physical effects qualitatively different to those found in static devices. This thesis has contributed to this effort along three main axes: improvements in numerical algorithms, improvements in software tools, and new experimental proposals.

The first contribution involved taking the existing formalism developed in Ref. [1] and improving it in two significant ways. The first improvement was the construction of an algorithm that scales as $\mathcal{O}(Nt_{max})$—where N is the number of degrees of freedom in the central system, and t_{max} is the maximum simulation time—as opposed to the best effort of Ref. [1], which has $\mathcal{O}(Nt_{max}^2)$ scaling. This was achieved by attaching slices of the leads to the central system and incorporating non-Hermitian terms into the lead Hamiltonian. The number of attached slices is tied to the required *accuracy* of the calculation, rather than being proportional to t_{max} as in Ref. [1]. Secondly, the integration over the initially filled states was moved from the energy domain to the momentum domain, which regularises the singularities associated with new modes opening in the leads.

The second contribution is an extension to the KWANTquantum transport package, called TKWANT, that can handle time-dependent problems. Although the software is not yet of production quality, it has clearly already added enormous value not only to our work, but also the work of collaborators. In addition there is a clear set of steps that are required to bring TKWANTto a standard where it can be released publicly alongside KWANT.

Using TKWANTwe then investigated the propagation of a charge pulse inside a flying qubit interferometer. The concept of dynamical control of interference, recently developed in Ref. [2], allowed us to interpret our results. The presence of this effect in this particular experimental setup bodes well for an experimental verification of this dynamical control in the near future. We then turned to superconducting systems and studied a Josephson junction under the action of static and time-varying

© Springer International Publishing AG 2017
J. Weston, *Numerical Methods for Time-Resolved Quantum Nanoelectronics*,
Springer Theses, DOI 10.1007/978-3-319-63691-7_8

bias. We were able to achieve *quantitative agreement* between our calculation for the sub-gap current-voltage characteristic of a short Josephson junction, and that obtained using purely analytical methods based on Floquet theory. This indicates that our time-resolved techniques are useful even for problems of a.c. transport that would traditionally be treated using Floquet theory. The $\mathcal{O}(Nt_{max})$ scaling of our numerical method really came into its own when investigating the propagation of charge pulses inside long Josephson junctions, where the infinite lifetime of the Andreev bound states means that extremely long simulation times are needed. Finally we combined the concepts of electronic interferometers and superconductivity to investigate the effect of voltage pulses on normal-insulator-normal-superconductor junctions in nanowires, which exhibit Majorana states. Understanding such a junction as a Fabry-Perot interferometer, we were able to see that repeated application of voltage pulses *stabilises the dynamical modification of the interference pattern*. This allowed us to perform "spectroscopy" on the Majorana states, which provides a signature of their nature as resonant Andreev states.

Future Perspectives

Time-resolved quantum electronics is still an emerging field and one that will, we think, see a heyday in the coming years. In particular, the use of single-electron sources in a new generation of experiments is a particularly exciting prospect. In this endeavour we can envisage two main axes for development in terms of numerical tools. Firstly, TKWANT (the software) needs to be brought to the stage where it is just as easy to set up and test new ideas for time-resolved transport as it is to use KWANT for stationary transport. This will allow theorists to more rapidly propose interesting experiments and help them gain an intuitive understanding of a new physical system before they bring analytical techniques to bear. Secondly, in order to obtain greater parity with experiment, TKWANT (the algorithm) should be modified to allow for a self-consistent calculation of observables that can be fed back into the Hamiltonian. The most obvious example would be to self-consistently solve the Poisson equation in order to account for electron-electron interactions on the mean-field level. Another interesting possibility would be to embed a Josephson junction into its surrounding (classical) circuit. This would allow us to go beyond simple models (such as RCSJ) by treating the full quantum dynamics of the Josephson junction. There are significant challenges to overcome, however, as the self-consistency effectively couples the wavefunctions at different energies, rendering the problem non-linear and making an efficient parallel implementation much more difficult. Despite the challenges, such developments would bring with them a wealth of new possibilities and, ultimately, new physics.

References

1. B. Gaury et al., Numerical simulations of time-resolved quantum electronics. Phys. Rep. **534**(1), 1–37 (2014)
2. B. Gaury, X. Waintal, Dynamical control of interference using voltage pulses in the quantum regime. Nat. Commun. **5**, 3844 (2014)

Appendix A
Discretising Continuous Models

This thesis deals primarily with performing numerical simulations of tight binding models. In several instances the model that we actually want to simulate is a continuous one, and a tight binding model is obtained by a discretisation procedure. In this section we shall show how in practice one obtains a discrete tight binding Hamiltonian from a continuous one.

We shall start from a general Hamiltonian of the form:

$$\hat{H} = \int_\Omega d\mathbf{x}\, \hat{\psi}^\dagger(\mathbf{x})\, H(\mathbf{x}, i\hbar\nabla, t)\, \hat{\psi}(\mathbf{x}) \tag{A.1}$$

where $\hat{\psi}^\dagger(\mathbf{x})$ ($\hat{\psi}(\mathbf{x})$) is a fermionic creation operator for a particle at position $\mathbf{x}$, $H(\mathbf{x}, i\hbar\nabla, t)$ is a differential operator—the realspace projection of the Hamiltonian—and Ω is the domain of the problem.

First we shall approximate the integral using a simple rectangle method over a uniform grid. Equation (A.1) is thus approximated by

$$\hat{H} \approx \sum_i \hat{\psi}^\dagger(\mathbf{x}_i)\, H(\mathbf{x}_i, i\hbar\nabla, t)\, \hat{\psi}(\mathbf{x}_i). \tag{A.2}$$

For $\Omega = \mathbb{R}^3$ we have $\mathbf{x}_i = a(n_i\mathbf{i} + m_i\mathbf{j} + l_i\mathbf{k})$, with a the discretisation step, $n_i, m_i, l_i \in \mathbb{Z}$, and $\mathbf{i}, \mathbf{j}$ and $\mathbf{k}$ are unit vectors in the x, y, and z directions respectively. For other Ω the allowed values of the n_i, m_i, l_i should be adjusted accordingly.

A.1 Finite Difference Formulae

Now we must approximate $H(\mathbf{x}, i\hbar\nabla, t)$, which we shall do using central finite-difference formulae. All the Hamiltonians dealt with in this thesis will be either linear or quadratic in $i\hbar\nabla$, so it will be enough to write down finite difference formulae

© Springer International Publishing AG 2017
J. Weston, *Numerical Methods for Time-Resolved Quantum Nanoelectronics*,
Springer Theses, DOI 10.1007/978-3-319-63691-7

for ∇ and ∇^2. We define δ_a, the finite difference operator with finite step a, in the following manner:

$$\nabla f(\mathbf{x}) \approx \delta_a[f](x) \equiv \frac{1}{2a} \sum_{\mathbf{n}} [f(\mathbf{x}+a\mathbf{n}) - f(\mathbf{x}-a\mathbf{n})]\,\mathbf{n}, \qquad (A.3)$$

where the sum runs over $\mathbf{n} \in \{\mathbf{i}, \mathbf{j}, \mathbf{k}\}$. While this expression has been written for three dimensions the equivalent expressions for one and two dimensions are obvious. We can obtain the finite difference operator for second derivatives by applying Eq. (A.3) twice:

$$\nabla^2 f(\mathbf{x}) \approx \delta_a^2[f](x) \equiv \frac{1}{a^2}[-6f(\mathbf{x}) + \sum_{\mathbf{n}} f(\mathbf{x}+a\mathbf{n}) + f(\mathbf{x}-a\mathbf{n})]. \qquad (A.4)$$

The factor of 6 in the term proportional to $f(\mathbf{x})$) is particular to 3D; in general there is a term $-2f(\mathbf{x})$ for each dimension. In evaluating these formulae we see that there are two types of term that appear: "onsite" terms, proportional to $f(\mathbf{x})$, and "hopping" terms proportional to $f(\mathbf{x} \pm a\mathbf{n})$. Note that in evaluating the formula for the second derivative we twice-applied the finite difference operator with discretisation spacing $a/2$, which in the end produced a scheme that only contains function evaluations at points spaced by a. This is advantageous for treating Hamiltonians that contain both first and second derivatives, as both only give first-nearest neighbour terms.

The above finite difference formulae are accurate to $\mathcal{O}(a^2)$ and only contain nearest-neighbour terms, but we could also define higher-order formulae accurate to $\mathcal{O}(a^4)$ or $\mathcal{O}(a^6)$ at the expense of having to include second and third nearest-neighbour terms, which renders the tight-binding Hamiltonian matrix less sparse. In this thesis we use the above $\mathcal{O}(a^2)$ accurate formulae exclusively.

A.2 From Continuum to Tight-Binding Models

If we apply the central difference formulae to the $\hat{\psi}(\mathbf{x}_i)$ of Eq. (A.2) we can see that we are going to generate terms at neighbouring points $\hat{\psi}(\mathbf{x}_j)$. The Hamiltonian will thus mix $\hat{\psi}^\dagger(\mathbf{x}_i)$ and $\hat{\psi}(\mathbf{x}_j)$ at neighbouring points. The most general form we can write down for Eq. (A.2) is then

$$\sum_{ij} H_{ij}\,\hat{\psi}^\dagger(\mathbf{x}_i)\,\hat{\psi}(\mathbf{x}_j), \qquad (A.5)$$

where the H_{ij} contains the coefficients from the finite difference formulae that bring the point $\mathbf{x}_j$ to $\mathbf{x}_i$. We recognise Eq. (A.5) as a tight-binding Hamiltonian in second quantisation.

Up till now this may seem rather abstract, so let us apply this to a concrete example: a particle in 1D in the presence of a potential landscape. We write the continuum Hamiltonian as

$$\hat{H} = \int_{-\infty}^{\infty} dx \, \hat{\psi}^{\dagger}(x) \left[\frac{-\hbar^2}{2m^2} \frac{\partial^2}{\partial x^2} + V(x,t) \right] \hat{\psi}(x) \tag{A.6}$$

and discretise onto the lattice of points $\{x_n = na \mid n \in \mathbb{Z}\}$ using the above procedure to obtain

$$\hat{H} \approx \sum_n \hat{\psi}^{\dagger}(x_n) \left[\frac{-\hbar^2}{2m^*a^2} (\hat{\psi}(x_{n+1}) + \hat{\psi}(x_{n-1}) - 2\hat{\psi}(x_n)) + V(x_n, t)\hat{\psi}(x_n) \right]. \tag{A.7}$$

Rearranging terms we get

$$\hat{H} \approx \sum_n \left[\left(\frac{\hbar^2}{m^*a^2} + V(x_n, t) \right) \hat{\psi}^{\dagger}(x_n)\hat{\psi}(x_n) \right.$$
$$\left. + \frac{-\hbar^2}{2m^*a^2} \hat{\psi}^{\dagger}(x_n)\hat{\psi}(x_{n+1}) + \frac{-\hbar^2}{2m^*a^2} \hat{\psi}^{\dagger}(x_n)\hat{\psi}(x_{n-1}) \right], \tag{A.8}$$

which allows us to identify the coefficients H_{ij} of the tight-binding model:

$$H_{ii} = \frac{\hbar^2}{m^*a^2} + V(x_i, t),$$
$$H_{i\,i+1} = -\frac{\hbar^2}{2m^*a^2}, \tag{A.9}$$

and

$$H_{i\,i-1} = -\frac{\hbar^2}{2m^*a^2}, \tag{A.10}$$

with all other terms 0.

Appendix B
The Peierls Substitution

Here we shall derive the so-called Peierls substitution [1, 2] that is commonly used to introduce a vector potential into tight-binding models. Contrary to most presentations in the literature we shall start from a continuum Hamiltonian, as all the models presented in this thesis start from such a description and arrive at a tight-binding Hamiltonian only after discretisation using the procedure outlined in Appendix A.1.

B.1 Peierls Substitution Without Magnetic Field

We shall first treat the case when there is no magnetic field: $\nabla \times \mathbf{A} = 0$. While this might seem contrived, we often treat cases where a time-dependent (but spatially uniform) voltage is applied to the leads of a nanoelectronic device. It is very useful to be able to express such a system in a gauge where there is a time-dependent vector potential *at the voltage drop*, as opposed to a time-dependent scalar potential everywhere in the lead. In this gauge the electric field is given by $\mathbf{E} = -\partial \mathbf{A}/\partial t$. This treatment allows us to identify the formal unitary gauge transformation presented in Appendix C.1 as the discrete analogue of the continuum *electromagnetic* gauge transformation presented here.

Let us start with a minimally-coupled Hamiltonian in first quantisation of the form

$$H(\mathbf{x}, i\hbar\nabla) = \frac{1}{2m}[-i\hbar\nabla + q\mathbf{A}(\mathbf{x})]^2. \tag{B.1}$$

Although we could try to discretise this Hamiltonian directly using the approach demonstrated in Appendix A, for the present case with no magnetic field it is actually simpler to first recast the Hamiltonian into a different form. Consider the action of the canonical momentum operator Π on a function $f(\mathbf{x})$:

$$\Pi f(\mathbf{x}) \equiv [-i\hbar\nabla + q\mathbf{A}(\mathbf{x})] f(\mathbf{x}). \tag{B.2}$$

© Springer International Publishing AG 2017
J. Weston, *Numerical Methods for Time-Resolved Quantum Nanoelectronics*,
Springer Theses, DOI 10.1007/978-3-319-63691-7

Compare this to the action of the operator $\mathbf{Q}$ on $f(\mathbf{x})$:

$$\mathbf{Q} f(\mathbf{x}) \equiv -i\hbar e^{-i(q/\hbar)\int_0^{\mathbf{x}} \mathbf{A}(\mathbf{x}') \cdot d\mathbf{x}'} \nabla \left(e^{i(q/\hbar)\int_0^{\mathbf{x}} \mathbf{A}(\mathbf{x}') \cdot d\mathbf{x}'} f(\mathbf{x}) \right), \qquad (B.3)$$

which we can evaluate explicitly—using the fact that $\int_0^{\mathbf{x}} \mathbf{A}(\mathbf{x}', t) \cdot d\mathbf{x}'$ is path independent ($\nabla \times \mathbf{A} = 0$) and hence a well-defined function of $\mathbf{x}$—to obtain

$$\mathbf{Q} f(\mathbf{x}) = [-i\hbar\nabla + q\mathbf{A}(\mathbf{x})] f(\mathbf{x}). \qquad (B.4)$$

We thus see that $\mathbf{Q} = \Pi$. We see that we can thus write the action of the Hamiltonian on $f(\mathbf{x})$ as

$$H(\mathbf{x}, i\hbar\nabla) f(\mathbf{x}) = \frac{-\hbar^2}{2m} e^{-i(q/\hbar)\int_0^{\mathbf{x}} \mathbf{A}(\mathbf{x}') \cdot d\mathbf{x}'} \nabla^2 \left(e^{i(q/\hbar)\int_0^{\mathbf{x}} \mathbf{A}(\mathbf{x}') \cdot d\mathbf{x}'} f(\mathbf{x}) \right). \qquad (B.5)$$

Using Eq. (A.4) we can write a discretised approximation for $H(\mathbf{x}, i\hbar\nabla) f(\mathbf{x})$:

$$H(\mathbf{x}, i\hbar\nabla) f(\mathbf{x}) \approx \frac{-\hbar^2}{2ma^2} \left[-6f(\mathbf{x}) + \sum_{\mathbf{n}} e^{i\varphi_{\mathbf{x}}(a\mathbf{n})} f(\mathbf{x} + a\mathbf{n}) + e^{i\varphi_{\mathbf{x}}(-a\mathbf{n})} f(\mathbf{x} - a\mathbf{n}) \right] \tag{B.6}$$

where $\varphi_{\mathbf{x}}(\mathbf{y}) = (q/\hbar) \int_{\mathbf{x}}^{\mathbf{x}+\mathbf{y}} \mathbf{A}(\mathbf{x}') \cdot d\mathbf{x}'$. We see that Eq. (B.6) is the same as the tight binding model in the absence of $\mathbf{A}$, except that the hopping terms have picked up phase factors.

B.2 Peierls Substitution With Magnetic Field

Let us now treat the case *with* magnetic field. Due to the fact that $\int_0^{\mathbf{x}} \mathbf{A}(\mathbf{x}') \cdot d\mathbf{x}'$ is now path dependent we cannot use the same formal reasoning as before. In the end, however, we will end up with the same general form for the discretised model, but this time it will be an *approximation* that is only valid when the magnetic field varies slowly on the length scale a of the discretisation. We start as before from the minimally coupled Hamiltonian, however this time we explicitly expand out the terms:

$$H(\mathbf{x}, i\hbar\nabla) f(\mathbf{x}) = \frac{-\hbar^2}{2m}\nabla^2 f(\mathbf{x}) - \frac{i\hbar q}{2m}[f(\mathbf{x})\nabla \cdot \mathbf{A}(\mathbf{x}) + 2\mathbf{A}(\mathbf{x}) \cdot \nabla f(\mathbf{x})] + q^2 \mathbf{A}(\mathbf{x}) \cdot \mathbf{A}(\mathbf{x}). \tag{B.7}$$

We now discretise this using the techniques of Appendix A.1 onto a cubic lattice with spacing a:

$$H(\mathbf{x}, i\hbar\nabla) f(\mathbf{x}) \approx -\sum_{\mathbf{n}} \left[\frac{\hbar^2}{2ma^2} + \frac{i\hbar q}{2ma} \mathbf{A}(\mathbf{x}) \cdot \mathbf{n} \right] f(\mathbf{x} + a\mathbf{n})$$

$$-\sum_{\mathbf{n}} \left[\frac{\hbar^2}{2ma^2} - \frac{i\hbar q}{2ma} \mathbf{A}(\mathbf{x}) \cdot \mathbf{n} \right] f(\mathbf{x} - a\mathbf{n}) \qquad (B.8)$$

$$+ \left[\frac{3\hbar^2}{ma^2} + q^2 |\mathbf{A}(\mathbf{x})|^2 - \frac{i\hbar q}{4ma} \sum_{\mathbf{n}} [\mathbf{A}(\mathbf{x} + a\mathbf{n}) - \mathbf{A}(\mathbf{x} - a\mathbf{n})] \cdot \mathbf{n} \right] f(\mathbf{x}).$$

Now we will re-express the above in units of $\gamma = \hbar^2/2ma^2$ in order to more easily see the relative order in powers of a of the different terms:

$$H(\mathbf{x}, i\hbar\nabla) f(\mathbf{x}) \approx -\gamma \sum_{\mathbf{n}} \left[1 + \frac{iqa}{\hbar} \mathbf{A}(\mathbf{x}) \cdot \mathbf{n} \right] f(\mathbf{x} + a\mathbf{n})$$

$$-\gamma \sum_{\mathbf{n}} \left[1 - \frac{iqa}{\hbar} \mathbf{A}(\mathbf{x}) \cdot \mathbf{n} \right] f(\mathbf{x} - a\mathbf{n})$$

$$+ \gamma \left[6 + \frac{2ma^2 q^2}{\hbar^2} |\mathbf{A}(\mathbf{x})|^2 - \frac{iqa}{2\hbar} \sum_{\mathbf{n}} [\mathbf{A}(\mathbf{x} + a\mathbf{n}) - \mathbf{A}(\mathbf{x} - a\mathbf{n})] \cdot \mathbf{n} \right] f(\mathbf{x}).$$
$$(B.9)$$

Now we will make some approximations in order to proceed. The first approximation will be that the magnetic field is constant over the length scale a. The consequence of this is that $\mathbf{A}(\mathbf{x})$ varies *linearly* over a distance a, and $\mathbf{A}(\mathbf{x} + a\mathbf{n}) - \mathbf{A}(\mathbf{x} - a\mathbf{n}) \propto 2a$. Next, in order to be compatible with the $\mathcal{O}(a^2)$ approximations to the first and second derivatives, we should discard all terms of $\mathcal{O}(a^2)$. This leaves us with

$$H(\mathbf{x}, i\hbar\nabla) f(\mathbf{x}) \approx -\gamma \sum_{\mathbf{n}} \left[1 + \frac{iqa}{\hbar} \mathbf{A}(\mathbf{x}) \cdot \mathbf{n} \right] f(\mathbf{x} + a\mathbf{n})$$

$$-\gamma \sum_{\mathbf{n}} \left[1 - \frac{iqa}{\hbar} \mathbf{A}(\mathbf{x}) \cdot \mathbf{n} \right] f(\mathbf{x} - a\mathbf{n}) \qquad (B.10)$$

$$+ 6\gamma f(\mathbf{x}).$$

The final part is to notice that $e^{ix} = 1 + ix + \mathcal{O}(x^2)$, and since we have discarded all terms of $\mathcal{O}(a^2)$ we can write Eq. (B.10) as:

$$H(\mathbf{x}, i\hbar\nabla) f(\mathbf{x}) \approx \frac{-\hbar^2}{2ma^2} \left[6f(\mathbf{x}) + \sum_{\mathbf{n}} e^{i\tilde{\varphi}_{\mathbf{x}}(a\mathbf{n})} f(\mathbf{x} + a\mathbf{n}) + e^{-i\tilde{\varphi}_{\mathbf{x}}(a\mathbf{n})} f(\mathbf{x} - a\mathbf{n}) \right],$$
$$(B.11)$$

with $\tilde{\varphi}_{\mathbf{x}}(\mathbf{y}) = (q/\hbar)\mathbf{A}(\mathbf{x}) \cdot \mathbf{y}$. We see that this is the same expression as Eq. (B.6), except that $\tilde{\varphi}_{\mathbf{x}}(a\mathbf{n})$ has replaced $\varphi_{\mathbf{x}}(a\mathbf{n})$. The Peierls substitution is therefore also valid when there is a magnetic field, except in this case it is an *approximation* that is valid when the magnetic field is roughly constant on length scales a.

References

1. R. Peierls, Zur Theorie des diamagnetismus von leitungselektronen. Zeitschrift fur Physik **80**(11–12), 763–791 (1933)
2. M. Graf, P. Vogl, Electromagnetic fields and dielectric response in empirical tight-binding theory. Phys. Rev. B **51**(8), 4940–4949 (1995)

Appendix C
Gauge Transformations

In this section we shall look at some important gauge transformations that are used in this thesis to bring a problem with time dependence in the infinite *leads* of a nanoelectronic system to a problem with time dependence only in a finite region.

First we recall the expression for a general, time-dependent gauge transformation on an arbitrary Hamiltonian. We start from the Schrödinger equation

$$i\hbar \frac{\partial}{\partial t}|\psi(t)\rangle = \hat{H}(t)|\psi(t)\rangle. \tag{C.1}$$

and define

$$|\psi(t)\rangle = \hat{U}(t)|\psi'(t)\rangle, \tag{C.2}$$

where $\hat{U}(t)$ is a unitary operator (the gauge transformation of interest), and plug C.2 into C.1:

$$i\hbar\left(\frac{\partial}{\partial t}\hat{U}(t)\right)|\psi'(t)\rangle + i\hbar\hat{U}(t)\frac{\partial}{\partial t}|\psi'(t)\rangle = \hat{H}(t)\hat{U}(t)|\psi'(t)\rangle. \tag{C.3}$$

Now we multiply on the left by $\hat{U}^\dagger(t)$ to obtain

$$i\hbar\frac{\partial}{\partial t}|\psi'(t)\rangle = \left(\hat{U}^\dagger(t)\hat{H}(t)\hat{U}(t) - i\hbar\hat{U}^\dagger(t)\frac{\partial}{\partial t}\hat{U}(t)\right)|\psi'(t)\rangle, \tag{C.4}$$

which we recognise as a Schrödinger equation for $|\psi'(t)\rangle$ with a modified Hamiltonian

$$\hat{H}'(t) = \hat{U}^\dagger(t)\hat{H}(t)\hat{U}(t) - i\hbar\hat{U}^\dagger(t)\frac{\partial}{\partial t}\hat{U}(t). \tag{C.5}$$

We thus conclude that Eqs. (C.1) and (C.4) represent the same physical situation, and Eq. (C.5) is the transformed Hamiltonian subject to a time-dependent gauge transformation $\hat{U}(t)$.

© Springer International Publishing AG 2017
J. Weston, *Numerical Methods for Time-Resolved Quantum Nanoelectronics*,
Springer Theses, DOI 10.1007/978-3-319-63691-7

C.1   Gauge Transformations in Semi-Infinite Leads

Several times in the main text we make use of the fact that systems that we consider have infinite, periodic leads that are time invariant. Here we shall show that if we start from a system where a lead has a uniform, time-varying potential applied to it we can perform a gauge transformation to remove the time dependence from the lead and bring it into the coupling between the lead and the scattering region. We shall treat a class of systems consisting of a finite, scattering region, S, with an arbitrary quadratic Hamiltonian coupled to a semi-infinite electrode, L, with a uniform but time-dependent voltage $V(t)$ applied to it. The Hamiltonian is written as

$$\hat{H}(t) = \underbrace{\sum_{ij} H_{ij}^S(t)\hat{c}_i^\dagger \hat{c}_j}_{\hat{H}^S(t)} + \underbrace{\sum_{ij} H_{ij}^T(t)\hat{c}_i^\dagger \hat{d}_j + h.c.}_{\hat{H}^T(t)} + \underbrace{\sum_{ij} \left(H_{ij}^L + V(t)\delta_{ij}\right)\hat{d}_i^\dagger \hat{d}_j}_{\hat{H}^L(t)}$$

$$(C.6)$$

where $\hat{c}_i^\dagger$ ($\hat{c}_j$) are the creation (annihilation) operators for fermions in the scattering region, and $\hat{d}_i^\dagger$ ($\hat{d}_j$) are the corresponding operators in the leads. Note that, as mentioned above, the time-dependence in the lead is given entirely by $V(t)$. We now choose to apply a gauge transformation

$$\hat{W}(t) = \prod_i \exp\left[-(i/\hbar)\varphi(t)\hat{d}_i^\dagger \hat{d}_i\right], \qquad (C.7)$$

where

$$\varphi(t) = \int V(t)\,dt. \qquad (C.8)$$

We shall now use Eq. (C.5) with Eqs. (C.6) and (C.7) to obtain the transformed Hamiltonian. First evaluating $\hat{W}^\dagger(t)\hat{H}(t)\hat{W}(t)$, term by term we note that $\left[\hat{W}(t), \hat{H}^S(t)\right] = 0$ because $\left[\hat{c}_i^\dagger \hat{c}_j, \hat{d}_k^\dagger \hat{d}_k\right] = 0$. Similar reasoning leads us to $\left[\hat{W}(t), \hat{H}^L(t)\right] = 0$. The $\hat{W}^\dagger(t)\hat{H}^T(t)\hat{W}(t)$ term requires the use of the following operator identities:

$$e^{i\alpha\hat{g}^\dagger\hat{g}}\,\hat{g}^\dagger\,e^{-i\alpha\hat{g}^\dagger\hat{g}} = e^{i\alpha}\,\hat{g}^\dagger \qquad (C.9)$$

$$e^{i\alpha\hat{g}^\dagger\hat{g}}\,\hat{g}\,e^{-i\alpha\hat{g}^\dagger\hat{g}} = e^{-i\alpha}\,\hat{g}, \qquad (C.10)$$

where α is a complex number and $\hat{g}$ is a fermionic operator (satisfying anticommutation relation $\{\hat{g}, \hat{g}^\dagger\} = 1$, and yields

$$\hat{W}^\dagger(t)\hat{H}^T(t)\hat{W}(t) = \sum_{ij} H_{ij}^T(t)e^{-i\varphi(t)}\hat{c}_i^\dagger \hat{d}_j + h.c. \qquad (C.11)$$

Now evaluating $-i\hbar\hat{W}^\dagger(t)\frac{\partial}{\partial t}\hat{W}(t)$ we get

$$-i\hbar\hat{W}(t)\frac{\partial}{\partial t}\hat{W}(t) = -V(t)\sum_{ii}\hat{d}_i^\dagger\hat{d}_i. \tag{C.12}$$

Putting this all together allows us to write the transformed Hamiltonian as

$$\hat{H}'(t) = \sum_{ij}H_{ij}^S(t)\hat{c}_i^\dagger\hat{c}_j + \sum_{ij}H_{ij}^T(t)e^{-i\varphi(t)}\hat{c}_i^\dagger\hat{d}_j + h.c. + \sum_{ij}H_{ij}^L\hat{d}_i^\dagger\hat{d}_j, \tag{C.13}$$

where we see that the time dependence has been moved from the leads into the lead-system coupling term. This calculation generalises trivially to the case where there are several semi-infinite electrodes, each with their own uniform, time-dependent voltage. In addition it is interesting to note that this gauge transformation coincides with that of Appendix B if our tight-binding model is the discretisation of a continuum model. This gauge transformation is, in effect, a transformation from the Coulomb gauge ($\nabla \cdot \mathbf{A} = 0$) to the Lorentz gauge ($\nabla \cdot \mathbf{A} + (1/c^2)\partial\varphi/\partial t = 0$).

C.2 Gauge Transformations for Superconducting Leads

Here we shall look at the case of a system with a superconducting lead that has a time-dependent bias applied to it. We shall see that while the problems look similar in the Lorentz gauge, in the Coulomb gauge there is an extra time-dependence that appears in the anomalous terms $\hat{c}^\dagger\hat{c}^\dagger$ in the Hamiltonian.

We shall start with the Hamiltonian of a system with a superconducting lead that has a time-dependent bias applied to it. In the Lorentz gauge the Hamiltonian reads:

$$\begin{aligned}
\hat{H}'(t) = &\sum_{ij}\left(H_{ij}^S(t) - \mu\delta_{ij}\right)\hat{c}_i^\dagger\hat{c}_j + \sum_{ij}H_{ij}^T(t)e^{-i\varphi(t)}\hat{c}_i^\dagger\hat{d}_j + h.c. + \\
&\sum_{ij}\left(H_{ij}^L - \mu\delta_{ij}\right)\hat{d}_i^\dagger\hat{d}_j + \Delta_{ij}\hat{d}_i^\dagger\hat{d}_j + h.c.,
\end{aligned} \tag{C.14}$$

with $\varphi(t)$ defined by Eq. (C.8). This is just Eq. (C.13) with the Fermi level μ subtracted and an extra anomalous term (see Sect. 6.1 for details). We shall now apply the gauge transformation

$$\hat{W}(t) = \prod_i\exp\left[(i/\hbar)\varphi(t)\hat{d}_i^\dagger\hat{d}_i\right], \tag{C.15}$$

(the inverse of Eq. (C.7)) to bring us into the Coulomb gauge. The treatment is identical to Appendix C.1, except that now we have to evaluate terms of the form

$$\Delta_{ij}\hat{W}^\dagger(t)\hat{d}_i^\dagger\hat{d}_j^\dagger\hat{W}(t). \tag{C.16}$$

Using Eq. (C.9) we see that

$$\Delta_{ij}\hat{W}^\dagger(t)\hat{d}_j^\dagger\hat{d}_j^\dagger\hat{W}(t) = \Delta_{ij}e^{2i\varphi(t)}\hat{d}_i^\dagger\hat{d}_j^\dagger, \tag{C.17}$$

i.e. the anomalous terms responsible for superconductivity pick up time-varying phase factors. The Hamiltonian in the Coulomb gauge, $\hat{H}(t)$, is then

$$\hat{H}(t) = \sum_{ij}\left(H_{ij}^S(t) - \mu\delta_{ij}\right)\hat{c}_i^\dagger\hat{c}_j \ + \sum_{ij}H_{ij}^T(t)e^{-i\varphi(t)}\hat{c}_i^\dagger\hat{d}_j + h.c. + $$
$$\sum_{ij}\left(H_{ij}^L - [\mu - V(t)]\delta_{ij}\right)\hat{d}_i^\dagger\hat{d}_j + \Delta_{ij}e^{2i\varphi(t)}\hat{d}_i^\dagger\hat{d}_j^\dagger + h.c. \tag{C.18}$$

It is now clear that even if the bias voltage is constant the Hamiltonian for treating this system will always be time dependent due to the phase factors—this is true regardless of the gauge in which we try to treat the problem. This inherent time dependence is what gives rise to the a.c. Josephson effect.

Printed in the United States
By Bookmasters